International Conference on Theoretical and Experimental Studies in Nuclear Applications and Technology (TESNAT 2016)

EPJ Web of Conferences Volume 128 (2016)

Hatay, Turkey
28 - 30 April 2016

Editors:

Ismail Hakki Sarpun
Eyyup Tel
Abdullah Aydin
Abdullah Kaplan

ISBN: 978-1-5108-3283-1

Printed from e-media with permission by:

Curran Associates, Inc.
57 Morehouse Lane
Red Hook, NY 12571

proceedings.com

Some format issues inherent in the e-media version may also appear in this print version.

Printed by Curran Associates, Inc. (2017)

For additional information, please contact EDP Sciences – Web of Conferences at the address below.

EDP Sciences – Web of Conferences
17, Avenue du Hoggar
Parc d'Activité de Courtabœuf
BP 112
F-91944 Les Ulis Cedex A
France

Phone: +33 (0) 1 69 18 75 75
Fax: +33 (0) 1 69 28 84 91

contact-edps@webofconferences.org

Additional copies of this publication are available from:

Curran Associates, Inc.
57 Morehouse Lane
Red Hook, NY 12571 USA
Phone: 845-758-0400
Fax: 845-758-2633
Email: curran@proceedings.com
Web: www.proceedings.com

TABLE OF CONTENTS

EPJ Web of Conferences **128**, 00001 (2016) DOI: 10.1051/epjconf/201612800001
TESNAT 2016

TESNAT 2016

International Conference on Theoretical and Experimental Studies in Nuclear Applications and Technology

TESNAT conference series aim to discuss and compare all applicable methods as are being applied at present in nuclear physics. The problems faced in these fields at present are focused in the development of new methods and in the improving of existing techniques to achieve an understanding of existing experimental data and in predicting with high reliability new properties and processes. This conferences propose to bring together all these related communities with the goal of creating an enriching dialog across the disciplines. The 2016 conference had given an overview on the theoretical and experimental challenges in nuclear physics and applications. The main topics of the workshops are computational nuclear physics, medical physics, Monte Carlo applications in Nuclear Physics and other applications of nuclear physics.

TESNAT 2016 attracted about 150 participants and during the workshop 5 lectures, 70 plenary talks as well as 50 posters were presented. We would like to thank all the participants as well as the members of the International Scientific Committee, the Local Organizing Committee, the sponsors: TAEK (Turkish Atomic Energy Authority) and especially the hosting Mustafa Kemal University and Payas Mayor for the extremely warm atmosphere of the workshop. We would also like to thank Prof. Dr. Arjan KONING, Prof. Dr. Mohamed BELGAID, Prof. Dr. Saleh SULTANSOY, Prof. Dr. Ahmet BOZKURT and Prof. Dr. Aysel KAYIŞ TOPAKSU for their supports and leaderships.

We would also like to thank Prof.Dr. İsmail Hakkı KARAHAN and Dr. Muhittin ŞAHAN for their supports, Dr. Bayram DEMİR, Dr. Ahmet BÜLBÜL, Ercan YILDIZ, Tuğba TUTAR KABLAN and Selma ÖZARSLAN for their helps in organization and the staff of "European Physical Journal Web of Conferences" for their help with publishing the proceedings.

İsmail Hakkı SARPÜN, Eyyup TEL, Abdullah AYDIN, and Abdullah KAPLAN

Editors

edp sciences

EPJ Web of Conferences

Statement of Peer review

In submitting conference proceedings to EPJ Web of Conferences, I certify to the Publisher that I adhere to the **Policy on Publishing Integrity** of the journal in order to safeguard good scientific practice in publishing.

1. All articles have been subjected to peer review administered by the proceedings editors.
2. Reviews have been conducted by expert referees, who have been requested to provide unbiased and constructive comments aimed, whenever possible, at improving the work.
3. Proceedings editors have taken all reasonable steps to ensure the quality of the materials they publish and their decision to accept or reject a paper for publication has been based only on the merits of the work and the relevance to the journal.

Title, date and place of the conference

TESNAT 2016,
28-30 APRIL 2016
HATAY, TURKEY

Proceedings editor(s):

İSMAİL HAKKI SARPÜN
EYYUP TEL
ABDULLAH AYDIN
ABDULLAH KAPLAN

Date and editor's signature

20 OCTOBER 2016 21.10.2016

EPJ Web of Conferences **128**, 01001 (2016) DOI: 10.1051/epjconf/201612801001
TESNAT 2016

Calculations of (n,2n) reaction cross sections for ^{74,76,78,80,82}Se up to 20 MeV

Halide Şahan[1,a], *Muhittin* Şahan[1], *Eyyup* Tel[1]

[1]*Osmaniye Korkut Ata University, Faculty of Arts and Science, Department of Physics, Osmaniye, Turkey*

Abstract. In the present work, the excitation functions of (n,2n) reactions for five isotopes of selenium (74,76,78,80,82Se) are calculated using ALICE/ASH, EMPIRE-3.2.2, PCROSS, and TALYS 1.6 computer codes based on statistical model up to 20 MeV. The theoretical calculations provide information of the (n,2n) excitation functions with the increasing target neutron number of selenium element. The calculated cross-sections were compared with experimental data from EXFOR and also with the cross-sections estimated with semi empirical formula developed by Tet et al. (2008) [18]. Results show a reasonably good agreement between the calculations and the experimental data from literature.

1 Introduction

The theoretical calculation models are necessary to provide the estimation of the particle–induced reaction cross sections due to the experimental difficulty [1,2]. In past years the cross section of selenium isotopes (^{74,76,78,80,82,84}Se) around 14-15 MeV have been measured by many researches such as Hille and Münzer [3]; Minetti and Pasquarelli [4]; Casanova and Sanchez [5]; Hoang et al. [6] and Guozhu He [7].

Selenium, discovered by Berzelius in 1817, is a non-metallic chemical element with symbol Se. Selenium with 34 protons and having from 40 to 48 neutrons is rarely occurs in its elemental state in nature as pure or compounds. There are six naturally isotopes of Selenium (^{74,76,78,80,82,84}Se), five of which are stable (^{74,76,77,78,80}Se). Selenium occurs in minerals such as eucairite (CuAgSe), crooksite (CuThSe) and clausthalite (PbSe). Selenium element is used extensively in electronics and metal alloys such as the lead plates and also glass industry to give a red color to glasses and enamels. Selenium is an essential trace element for some species, including humans. Our bodies contain about 14 milligrams, and every cell in a human body contains more than a million selenium atoms. It can cause health problems in human body.

In the present work, neutron interactions with five selenium isotopes as ^{74}Se(n,$2n$)^{73}Se, ^{76}Se(n,$2n$)^{75}Se, ^{78}Se(n,$2n$)^{77}Se, ^{80}Se(n,$2n$)^{79}Se, and ^{82}Se(n,$2n$)^{81}Se by four nuclear reaction codes such as ALICE/ASH [8], EMPIRE [9], TALYS [10], and PCROSS [11] have been investigated and then compared with experimental data from EXFOR [12] in the literature and with evaluated data ENDF/B-VII.1 (2014) library.

The empirical formula for (n,$2n$) reaction cross sections at neutron energy of 14.5 MeV was developed by Tel et al. [13]. These formulas have been given as follows;

$$\ln \sigma_{n,2n} = \begin{Bmatrix} 7.15\left[1-2.45e^{-31.620(N-Z)/A}\right] & \textit{for even } A \\ 7.65\left[1-1.59e^{-23.06(N-Z)/A}\right] & \textit{for odd } A \end{Bmatrix} \qquad (1)$$

For (n,2n) reaction cross sections of ^{74}Se(n,$2n$)^{73}Se, ^{76}Se(n,$2n$)^{75}Se, ^{78}Se(n,$2n$)^{77}Se, ^{80}Se(n,$2n$)^{79}Se and ^{82}Se(n,$2n$)^{81}Se at 14-15 MeV, the calculated cross sections for isotopes of $^{74,76,78,80,82}Se$ were compared with a semi empirical formula developed by Tel et al. [13].

2 Results and Discussions

In this work, as target element, cross sections for ^{74}Se(n,$2n$)^{73}Se, ^{76}Se(n,$2n$)^{75}Se, ^{78}Se(n,$2n$)^{77}Se, ^{80}Se(n,$2n$)^{81} and ^{82}Se(n,$2n$)^{81} reactions have been theoretically calculated at neutron energy centred around 14-15 MeV using four computer codes based on statistical model. These computer codes are ALICE/ASH [8], EMPIRE-3.2.2 [9], PCROSS [11], and TALYS 1.6 [10]. These model calculations were compared with the semi empirical formula developed by Tel et al. [13] and experimental data form EXFOR.

Figs. 1-5 show theoretical calculations of ALICE/ASH, EMPIRE, PCROSS, and TALYS 1.6, experimental data form EXFOR, semi-empirical data from Tet et al. (2008), and evaluated ENDF/B-VII.1 data for the $^{74,76,78,80,82}Se$ reactions. More *detailed* information about our calculations and comparisons for the $^{74,76,78,80,82}Se$ reactions are given below.

Calculated cross sections of $^{74}Se(n,2n)$ ^{73}Se reaction between 10 and 20 MeV is shown in Fig. 1. The experimental data obtained from six different researches using EXFOR at 10 to 20 MeV and the empirical cross section data evaluated by Tel et al. [13] were compared

[a]Corresponding author: muhittinsahan@osmaniye.edu.tr

with PCROSS, EMPIRE and TALYS and ALICE/ASH data. While the cross section data from TALYS 1.6 code has higher value to be 441.0 mb, which is approximately the same value of ENDF/B-VII.1 data, the cross section data from PCROSS has lower value to be 223.14 mb as seen from Fig. 1. All calculation data in Fig. 1 almost agree well with the measured data from EXFOR at 13 MeV up to 17 MeV. Semi-empirical cross section data evaluated by Tel et al. [13] estimated to be ~330.60 mb at 14.5 MeV is closer agreement with the data calculated using PCROSS, EMPIRE, and ALICE/ASH data.

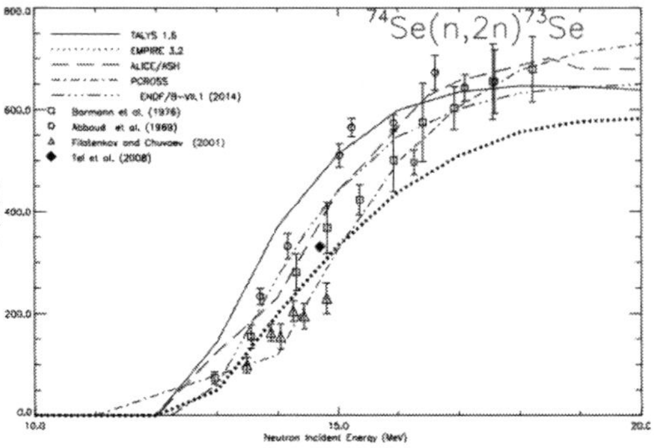

Fig. 1. Excitation function of ^{74}Se(n,$2n$)^{73}Se reaction cross sections calculated by TALYS, EMPIRE, ALICE/ASH, and PCROSS and ENDF/B-VII.1 (2014) along with experimental data and semi-empirical calculation.

The cross-section data of $^{76}Se(n,2n)^{75}Se$ reaction are given in Fig. 2. The theoretical calculations produced by EMPIRE, TALYS, PCROSS and ALICE/ASH are compared with the earlier six experimental data from EXFOR and are also compared with the empirical data (Tel, et al.) [13] at 14.5 MeV energy. In general, Fig. 2 illustrates a marked increase in energy value with increasing cross section of ^{76}Se(n,$2n$)^{75}Se for the empirical cross-section calculations and experimental data. Fig. 2 shows the degree of agreement between the theoretical and experimental cross sections.

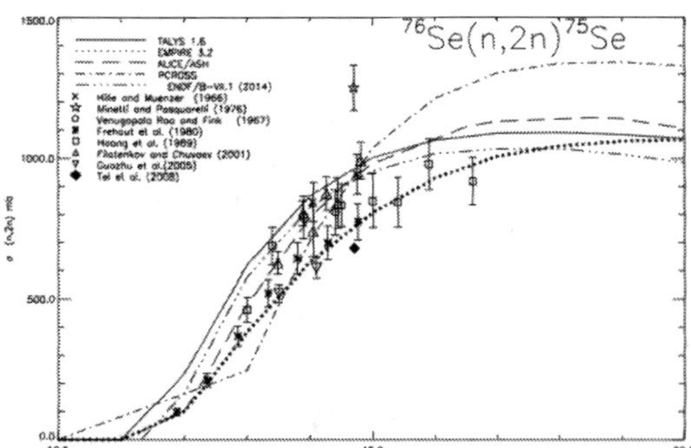

Fig. 2. Excitation function of ^{76}Se(n,$2n$)^{75}Se reaction cross sections calculated by TALYS, EMPIRE, ALICE/ASH, and PCROSS and ENDF/B-VII.1 (2014) along with experimental data and semi-empirical calculation.

Fig. 3 shows the excitation function of the nuclear reaction of ^{78}Se(n,$2n$)^{77}Se up to 20 MeV. Examining the nuclear reaction ^{78}Se(n,$2n$)^{77}Se in Fig. 3, the EMPIRE code calculation has provided a full harmony with experimental data of Frehaut et al. (1980) from EXFOR [12] 10 MeV and 15 MeV. Semi-empirical cross-section data developed by Tel et al. [13] has same cross section of 940 mb at 14-15 MeV.

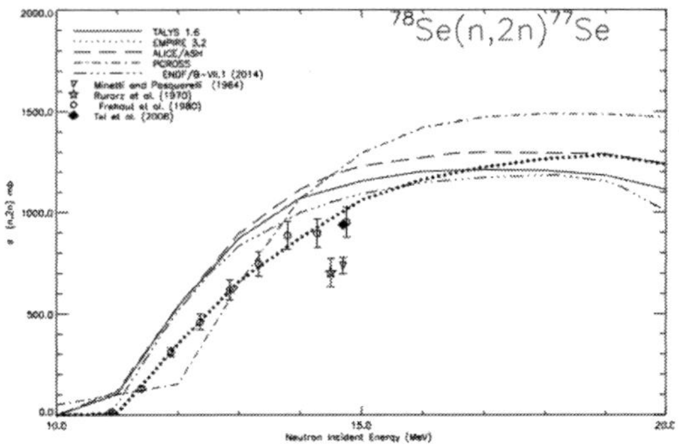

Fig. 3. Excitation function of ^{78}Se(n,$2n$)^{77}Se reaction cross sections calculated by TALYS, EMPIRE, ALICE/ASH, and PCROSS and ENDF/B-VII.1 (2014) along with experimental data and semi-empirical calculation.

The excitation function of the ^{80}Se(n,$2n$)^{79}Se nuclear reaction between 10 MeV and 20 MeV is shown in Fig. 4. Only one suitable experimental data with high error bars given by Bormann et al. (1967) from EXFOR [8] is available for ^{80}Se(n,$2n$)^{79}Se reaction in literature. There is a good correlation between theoretical calculations from four computer codes and all semi-empirical data at 14-15 MeV and experimental data. Semi-empirical cross-section data developed by Tel et al. [13] has same cross section of 940 mb at 14-15 MeV. EMPIRE from four models at 14-15 MeV is the same cross section value (mean 1100 mb) with semi-empirical cross-section data developed by Tel et al. [13]. Experimental data from EXFOR at 14-15 MeV have the same value with ALICE/ASH, TALYS and PCROSS codes. For ^{80}Se(n,$2n$)^{79}Se reactions, the experimental data are between 13.0- 16.75 MeV as seen from Fig. 4.

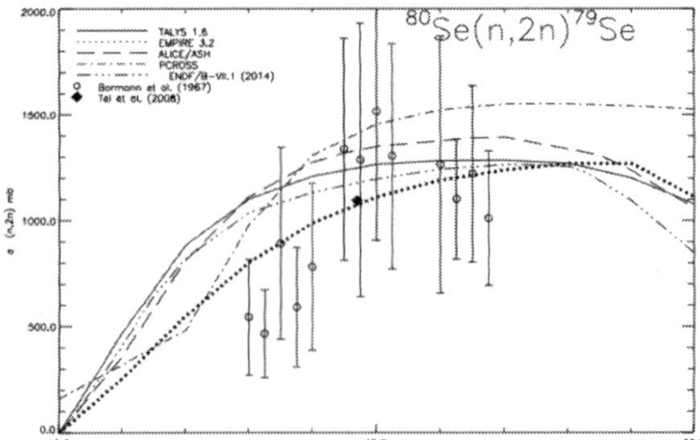

Fig. 4. Excitation function of ^{80}Se(n,$2n$)^{79}Se reaction cross sections calculated by TALYS, EMPIRE, ALICE/ASH, and

DOI: 10.1051/epjconf/201612801001

PCROSS and ENDF/B-VII.1 (2014) along with experimental data and semi-empirical calculation.

The computed cross sections based on ALICE/ASH, EMPIRE, PCROSS, and TALYS together with experimental data (EXFOR) [12] and evaluated ENDF/B-VII.1 data file for (n,2n) reaction on ^{82}Se isotope from 10 MeV to 20 MeV are shown in Fig. 5. Our theoretical values do agree with the available experimental data in the energy range of 10-15 MeV. Four experimental data are available for $^{82}Se(n,2n)^{81}Se$ reaction. The evaluated values of cross-sections are in reasonable agreement with the experimental values reported by Minetti and Pasquarelli (1967), Hasan et al. (1972), Kao and Alford (1975), and Filatenkov (2001) (All from EXFOR) [12].

3 Conclusion

In this work, cross sections for ^{74,76,78,80,82}Se reactions have been calculated in the energy range from 10 to 20 MeV using nuclear model calculations such as ALICE/ASH [8], TALYS 1.6 [10], EMPIRE-3.2.2 [9], and PCROSS [11]. In order to compare with the model calculations, we obtained ENDF/B-VII.1 results. As seen all figures, cross section values increase generally from 10 MeV to 15 MeV and then decrease to 20 MeV. The calculated cross sections are a reasonably good agreement between the calculated and experimental data, which is available in EXFOR database library.

References

1. R.A. Forrest, J. Kopecky, Fusion Engineering and Design, **82**, 73 (2007)
2. L. Goyal, P. Gur, Pramana-J. Phys.**72**, (2), 355 (2009)
3. P. Hille, H. Münzer, Acta Phys. Austr. **23**, 44 (1966)
4. B. Minetti, A. Pasquarelli, Nucl. Phys. **A100**, 186 (1967)
5. J.L. Casanova, M.L.Sanchez, Ann. Fis. (Spain) **72**, 186 (1976)
6. H.M. Hoang, U. Garuska, A. Marcinkowski, B. Zwieglinski, Z. Phys. A Atomic Nuclei **334**, 285-291 (1989)
7. He Guozhu, Liu Zhongjie, LuoJunhua, Kong Xiangzhong, Indian Journal of Pure & Applied Physics ,**43**, 729, (2005)
8. C.H.M. Broeders, A.Yu. Konobeyev, Yu.A. Korovin, V.P. Lunev, M. Blann, ALICE/ASH manual, FZK 7183, http://bibliothek.fzk.de/zb/berichte/FZKA7183.pdf (2006)
9. M. Herman, et al., EMPIRE-3.2 Malta modular system for nuclear reaction calculations and nuclear data evaluation, User's Manual (2013)
10. A. Koning, S. Hilaire, S. Goriely, TALYS 1.6- A Nuclear Reaction Program, User Manual, 1st edition (NRG, The Netherlands, 2013).
11. R. Capote, et al., Final Report on Research contract 5472/RB, INDC (CUB)-004, (Higher Institute of Nuclear Science and Technology, Cuba). Translated by the IAEA, (PCROSS program code) (1991)
12. EXFOR/CSISRS, Brookhaven National Laboratory, National Nuclear Data Center, (http://www.nndc.bnl.gov/exfor/) (2009)
13. E. Tel, Ş. Okuducu, M.H. Bölükdemir, G. Tanir, Int. J. Mod. Phys. E, **17** (3), 567 (2008)

 DOI: 10.1051/epjconf/201612801002

Theoretical cross sections of tantalum on neutron induced reactions

Tarik Siddik

Salahaddin University-College of Basic Education-Department of General Science, Erbil, Iraq

Abstract. Neutron-induced cross-sections for the stable isotope ^{181}Ta, in the energy region up to 20 MeV have been calculated. Statistical model calculations, based on the Hauser–Feshbach formalism, have been carried out using the TALYS-1.0 and were compared with available experimental data in the literature and with ENDF/B-VII, T = 300 K; JENDL-3.3, T = 300 K and JEFF-3.1, T = 300 K evaluated libraries.

1 Introduction

With the advent of fast computers, software that simulates nuclear reactions is able to play an increasingly important role in nuclear data. It is possible to provide an exact computational scheme for sophisticated nuclear models, not only new ones but also those that have been lying on the shelf for decades and only now become amenable for numerical implementation. Large scale comparisons with measurements are within reach. TALYS is a nuclear reaction program created at NRG Petten, the Netherlands, and CEA Bruyères-le-Châtel,France. The purpose of TALYS is to simulate nuclear reactions that involve neutrons, photons, protons, deuterons, tritons, ^{3}He, and α particles in the 1 keV to 200 MeV energy range. Predicted quantities include integrated, single and double-differential cross sections, for both the continuum and discrete states, residue production and fission cross sections, γ -ray production cross sections, etc. Nuclear cross section data can be obtained by three different ways on the basis of experimental measurement, theoretical model calculations and evaluation data files. Experimental cross section data in the range 15-20 MeV are not available for all neutron induced reactions on Tantalum isotopes however, there is only a limited number of data for at the energy range <15 MeV [1- 4].

2 Calculation Methods

The high-energy end of the ejectile spectra are described by preequilibrium emission, which takes place after the first stage of the reaction but long before statistical equilibrium of the compound nucleus is attained. It is imagined that the incident particle creates step by step more complex states in the compound system and gradually loses its memory of the initial energy and direction. The default preequilibrium model of TALYS is the two-component exciton model [5,6]. We recall the basic formula of [5] for the exciton model cross section,

$$\frac{d\sigma_k^{\rm EM}}{dE_k} = \sigma^{\rm CF} \sum_{p_\pi=p_\pi^0}^{p_\pi^{\rm eq}} \sum_{p_\nu=p_\nu^0}^{p_\nu^{\rm eq}} w_k(p_\pi, h_\pi, p_\nu, h_\nu, E_k) \times S_{\rm pre}(p_\pi, h_\pi, p_\nu, h_\nu), \quad (1)$$

where pπ (pν) is the proton (neutron) particle number and hπ (hν) the proton (neutron) hole number, $\sigma^{\rm CF}$ is the compound formation cross section, and $S_{\rm pre}$ is the time-integrated strength which determines how long the system remains in a certain exciton configuration. The initial proton and neutron particle numbers are denoted $p_\pi^0 = Z_p$ and $p_\nu^0 = N_p$ with $Z_p(N_p)$ being the proton (neutron) number of the projectile. In general, $h_\pi = p_\pi$ and $h_\nu = p_\nu - p_\nu^0$ so that the initial hole numbers are zero, i.e., $h_\pi^0 = h_\nu^0 = 0$ for primary preequilibrium emission. The preequilibrium part is calculated by Eq. (1), using $p_\pi^{eq} = p_\nu^{eq} = 6$, whereas the remainder of the reaction flux is distributed through the Hauser-Feshbach model. In addition, the never-come-back approximation is adopted. The emission rate wk for ejectile k with spin sk is given by

$$w_k(p_\pi, h_\pi, p_\nu, h_\nu, E_k) = \frac{2s_k+1}{\pi^2\hbar^3}\mu_k E_k \sigma_{k,\rm inv}(E_k) \times \frac{\omega(p_\pi - Z_k, h_\pi, p_\nu - N_k, h_\nu, E_x)}{\omega(p_\pi, h_\pi, p_\nu, h_\nu, E^{\rm tot})}, \quad (2)$$

where $\sigma_{k,\rm inv}$(Ek) is the inverse reaction cross section as calculated from the optical model, and ω is the two-component particle-hole state density. The full reaction dynamics that leads to Eq. (1) is described in Refs. [5,6]. We here restrict ourselves to the formulas given above since they contain the model- and parameter-dependent quantities. The expression for $S_{\rm pre}$ contains the adjustable transition matrix element M2 for each possible transition between neutron-proton exciton

DOI: 10.1051/epjconf/201612801002

configurations. A proton-neutron ratio of 1.6 for the squared internal transition matrix elements was adopted to give the best overall agreement with experiment, i.e., $M_{\pi\nu}^2 = M_{\nu\pi}^2 = 1.6M_{\nu\pi}^2 = 1.6M^2$. Partial level density parameters gπ=Z/17 and gν= N/17 were used in the equidistant spacing model for the partial level densities. Finally, an effective surface interaction well depth V=12 MeV [5] was used. At incident energies above several tens of MeV, the residual nuclides formed after binary emission may have so large an excitation energy that the presence of additional fast particles inside the nucleus becomes possible. The latter can be imagined as strongly excited particle-hole pairs resulting from the first binary interaction with the projectile. The residual system is then clearly non-equilibrated, and the excited particle that is high in the continuum may, in addition to the first emitted particle, be emitted on a short time scale. This so-called multiple preequilibrium emission forms an alternative theoretical picture of the intranuclear cascade process, whereby the exact location and momentum of the particles are not followed, but instead the total energy of the system and the number of particle-hole excitations (exciton number). In actual calculations, the particle-hole configuration of the residual nucleus after emission of the ejectile, is reentered as an initial condition in Eq. (1). When looping over all possible residual configurations, the multiple preequilibrium contribution is obtained [7-20]. In TALYS, multiple preequilibrium emission is followed up to arbitrary order; though for 96 MeV, only the secondary preequilibrium emission is significant. It is well known that semiclassical models, such as the exciton model, have had some problems in describing angular distributions (essentially because the model is based on a compound like concept instead of a direct one). Therefore, as mentioned previously, the double-differential cross sections are obtained from the calculated energy spectra using the Kalbach systematics [11]. To account for the evaporation peaks in the charged-particle spectra, multiple compound emission was treated with the Hauser-Feshbach model. In this scheme, all reaction chains are followed until all emission channels are closed. The Ignatyuk model [12] has been adopted for the total level density to account for the damping of shell effects at high excitation energies. For preequilibrium reactions involving deuterons, tritons, ^{3}He and α particles, a statistical contribution from the exciton model is automatically calculated with the formalism described above. However, it is well known that for nuclear reactions involving projectiles and ejectiles with different particle numbers, mechanisms such as stripping, pickup, and knockout play an important role, and these direct like reactions to the continuum are not covered by the exciton model. Therefore, Kalbach has developed a phenomenological contribution for these mechanisms [11], which is included in TALYS. The advantages over the older method include a better consideration of the available phase space through normalized particle-hole state densities and a better empirical determination of the pickup, stripping, knockout strength parameters, enabled by the more extensive experimental database that is now available. It has recently been shown that for medium and heavy nuclides this method gives a considerable improvement over the older methods. The latter seemed to consistently under predict neutron-induced reaction cross sections involving pickup of one or a few nucleons.

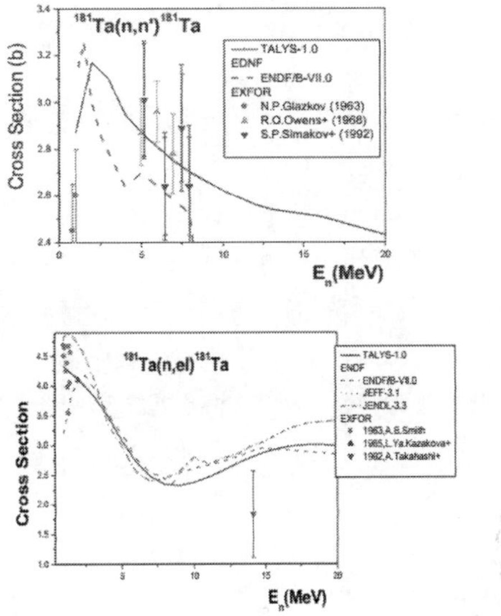

Fig. 1. The comparison of calculated excitation function using TALYS of ^{181}Ta(n,n') & ^{181}Ta(n, el) reaction with available experimental values and evaluated nuclear data files ENDF/B-VI . The values reported in Ref. (21)

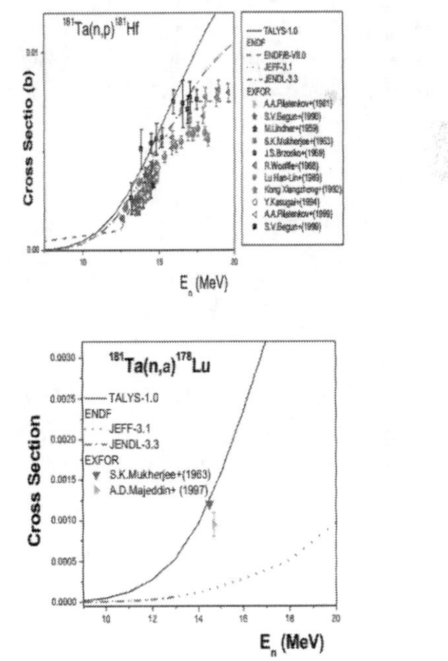

Fig. 2. The comparison of calculated excitation function using TALYS of ^{181}Ta(n,p) and ^{181}Ta(n,α) reaction with available experimental values and evaluated nuclear data files ENDF/B-VI, JENDL-3.3 and JEFF3.1.The values reported in Ref. (21)

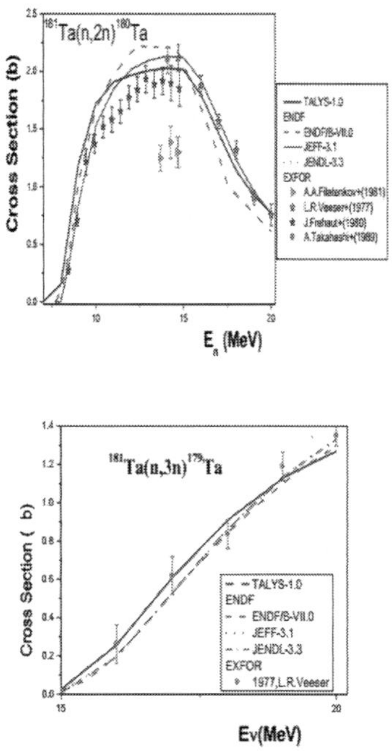

Fig. 3. The comparison of calculated excitation function using TALYS of ^{181}Ta(n,2n) & ^{181}Ta(n,3n) reaction with available experimental values and evaluated nuclear data files ENDF/B-VI, JENDL-3.3 and JEFF3.1. The values reported in Ref. (21)

3 Calculations and analysis

In Figs. 1–3, ^{181}Ta(n,inl), ^{181}Ta(n,el), ^{181}Ta(n,p), ^{181}Ta(n,α), ^{181}Ta(n,2n), ^{181}Ta(n,3n) reaction cross sections have been shown, respectively. The new calculations on the excitation functions of present paper have been carried out up to 20 MeV incident neutron energy. The results of our calculations using TALYS-1.0 code are compared with available experimental data in the EXFOR/CSISRS (2008) and with ENDF/B-VII.0 T = 300 K, JENDL-3.3 T = 300 K and JEFF3.1, T=300k evaluated data libraries. A reasonable agreement with experimental and theoretical excitation functions was obtained. The results can be summarized and conclude as follows:

3.1 (n,inl) Reactions

There are only a limited number of experimental cross section data for ^{181}Ta (n,inl) reaction in the energy range up to 20 MeV Fig. 1. The calculation of TALYS-1.0 code of ^{181}Ta (n,inl) reaction are in good agreement with the measurements between 5 and 8 MeV.

3.2 (n,el) Reactions

There are only a limited number of experimental cross section data for ^{181}Ta (n,inl) reaction in the energy range up to 20 MeV Fig. 1.

3.3 (n,p) Reactions

High discrepancies are observed in experimental data for ^{181}Ta(n,p) The calculation of TALYS-1.0 code of ^{181}Ta(n,p) reaction are in good agreement with the measurements between 12 and 17MeV Fig. 2. The evaluated nuclear data files JEFF-3.1 and JENDL-3.3 Show important discrepancies in energies < 16 MeV.

3.4 (n,α) Reactions

There are only two experiment value for ^{181}Ta(n,α) Fig. 2. TALYS-1.0 has been used for ^{181}Ta(n,α) reaction.

3.5 (n,2n) Reactions

The calculation of TALYS-1.0 code of ^{181}Ta(n,2n) reaction are in good agreement with the measurements between 12 and 20 MeV Fig. 3. ENDEF/B-VI, JENDL-3.3 and JEFF-3.1 files are in agreement with each other.

3.6 (n,3n) reactions

The calculation of TALYS-1.0 code of ^{181}Ta(n,3n) reaction are in good agreement with the experimental data Fig. 3. ENDEF/B-VI, JENDL-3.3 and JEFF-3.1 files are in agreement with each other.

References

1. T. Siddik, J. Fusion Energ., **32**, 164, (2013).
2. E. Tel, J. Fusion Energ., **29**, 332, (2010).
3. T. Siddik, J. Fusion Energ., **34**, (2015).
4. T. Siddik, *30th International Physics Congress/* Istanbul/Turkey (2013).
5. A.J. Koning, M.C. Duijvestijn, Nucl. Phys. **A744**, 15 (2004).
6. C. Kalbach, Phys. Rev., **C33**, 818 (1986).
7. T. Siddik, J. Advanced Physics, **5**, 1, (2016).
8. E. Tel, et al., Acta Phys. Slov. **54** (2), 191 (2004).
9. A.J. Koning, J.-P. Delaroche, Nucl. Phys. **A713**, 231 (2003).
10. A.J. Koning, et al., TALYS-1.0, *Proceedings of the International Conference on Nuclear Data for Science and Technology*, 22–27 May (2007)
11. C.Kalbach, Phys.Rev., **C37**, 2350(1989).
12. A.V. Ignatyuk, et al., Sov. J. Nul. Phys. **21**, 255 (1975).
13. A. Aydin, J. of Fusion Energy **29**(5), 476 (2010).
14. A. Kaplan, J. Fusion Energ. **32 (3)**, 382 (2013).
15. B. Sarer et al., Ann. Nucl. Energ. **45**(36), 417 (2009).
16. T. Belgya, et al., *Handbook for Calculations of Nuclear Reaction Data, RIPL-2* (IAEA-Tecdoc-1506), Taylor & Francis, (2006).
17. M. Yigit, J. Fusion Energ **34**:1392 (2015)
18. H. Korkut, et al., J. Fusion Energy, **35, 3,** 591 (2016).
19. M. Aikawa, et al., Astron. Astrophys. **441**, 1195, (2005).
20. E.Tel , A.Kara, J. Fusion Energy. **31, 3,** 257 (2012).
21. Brookhaven National Laboratory, National Nuclear Data Center, EXFOR/CSISRS (Experimental Nuclear Reaction Data File) Database Version of November 24 (2008).

EPJ Web of Conferences **128**, 01003 (2016) DOI: 10.1051/epjconf/201612801003
TESNAT 2016

Investigation of the excitation functions for some medical radioisotopes production

Fatma Kılınç[1,a], *Nurdan* Karpuz[2],and *Betül* Çetin[1]

[1] Amasya University, Department of Physics, Amasya, Turkey
[2] Amasya University, Sabuncuoğlu Şerefeddin Health Services Vocational School, Amasya, Turkey

Abstract. One of the main application fields of nuclear technology is medicine and radioisotopes are used in medicine. Production of those radioisotopes is important and in the production processes the cross section must be known. All the production of radioisotope used in medicine is based on the nuclear reactions means they are not natural. The decay time of produced radioisotopes is important as from production to hospital can take time and thus generally generator is used to produce some radioisotopes. Radioisotopes are widely produced in reactors or cyclotron type accelerator. Type of radioisotopes direct way to be used in production processes. Thus obtaining of cross section becomes crucial. For this purposes the theoretical calculation cross section of some radioisotopes used in medicine will be calculated in this study. The calculations will be done using Monte Carlo code of TALYS 1.6

1 Introduction

Nuclear physics, in the sense of the emergence of important technological innovation is a field of Science. Nowadays, it has a wide usage. These areas, which is one of nuclear medicine, are used for both diagnostic and treatment purposes and helps in diagnosis. Radionuclide generators can be produced with the desired properties by means of radioisotopes for diagnosis and treatment. If we look in terms of nuclear reactions the obtained experimental results important for the understanding of basic nuclear physics.

Energy of nuclear reactions energy dependence of the cross section and the spectral value is not known in detail for as numerous of delivery must be examined. Radioisotopes produced in reactors typically have short half-lived. The spectra of these nuclei cross-section and delivery is unlikely to be measured directly. In this sense, in terms of gaining time, we must do this first, theoretical nuclear reaction cross sections it is necessary to calculate in advance.

Experimental measurements of long-term, costly and does not occur at the same energy for different isotopes. Understanding the mechanisms of experimental and theoretical reaction cross section, there is a need to develop models to be used to. At the same time, verification of the theoretical model and the empirical formula of the studies, in order to create the experimental data is referenced. On the other hand, with the theoretical cross section predictions obtained with the nuclear models successful and scientifically verified data, may correct a deficiency that cannot be obtained in the experimental way. In this case, both experimental and theoretical studies of the cross section, there is a significant need in this area. In addition, the obtained nuclear data, nuclear structure and nuclear models for the investigation of theoretical studies has an important place in the development. [1].

2 Methods

^{13}N, ^{18}F are the radioisotopes used in nuclear medicine. Besides, ^{12}C, ^{20}Ne, ^{16}Oisotopes used in a medical sense, ^{13}N, ^{18}F is used as target material for the production of radioisotopes. These radioisotopes are used in research on the brain.

In this study, the cross section of ^{12}C(d,n), ^{20}Ne(d,α), ^{16}O(^{3}He,p) reactions was calculated. The calculations, it was theoretically with Talys1.6 program by Monte Carlo simulation code for nuclear reaction. TALYS is a nuclear reaction simulation computer code system for the analysis and prediction of nuclear reactions. Talys 1.6 [2] program 1keV-1GeV in the energy range of the proton, neutrons, deuteron, triton, gamma, alpha and ^{3}He particles with a mass of 12 and more, which examines the interaction of the core. TALYS nuclear reaction simulation programme to make predictions about how you can achieve results with different models for reaction cross section offers the possibility. The program has two main objectives. Firstly, experiments in the theoretically of nuclear reactions can be performed. Secondly, the nuclear data tool is needed. It has the FORTRAN programming language. Runs on Linux operating system.

[a]Corresponding author: fkilinc27@hotmail.com

2.1 Figures

The calculated $^{12}C(d,n)$, $^{20}Ne(d,\alpha)$, $^{16}O(^{3}He,p)$ reaction cross sections have been displayed in Figs. 1, 2 and 3.

The obtained results have been compared with the experimental data existing in the EXFOR [3] databases.

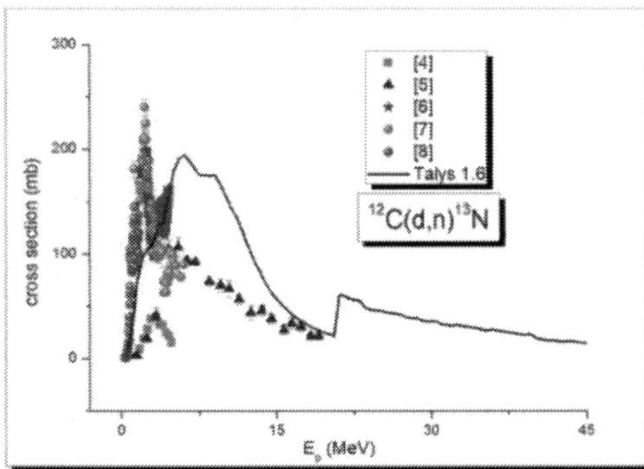

Fig. 1. The calculated cross sections of $^{12}C(d,n)$ reaction and comparison with experimental results.

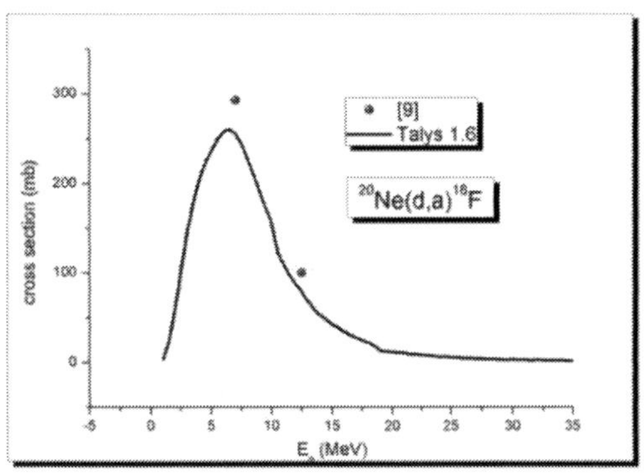

Fig. 2. The calculated cross sections of $^{20}Ne(d,\alpha)$ reaction and comparison with experimental results.

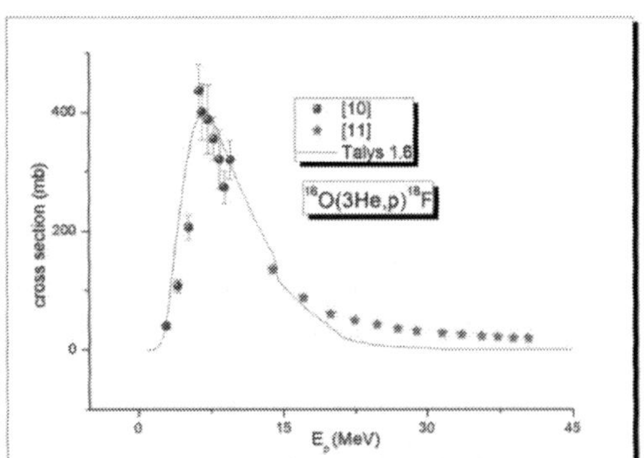

Fig 3. The calculated cross sections of $^{16}O(^{3}He,p)$ reaction and comparison with experimental results.

3 Conclusion

In this work, $^{12}C(d,n)$, $^{20}Ne(d,\alpha)$, $^{16}O(^{3}He,p)$ reaction cross section values for the reaction were calculated in the energy range between 0.5 and 100 MeV. Our results are in fair agreement with both, available experimental data in literature. The maximum cross section values are 194.913mb at 6 MeV at the $^{12}C(d,n)$ nuclear reaction and 260.383mb at 6.5 MeV at the $^{20}Ne(d,\alpha)$ and 399.559 mb at 7 MeV at the $^{16}O(^{3}He,p)$. ^{13}N is the most appropriate production range; for $^{12}C(d,n)$ nuclear reaction is seen that the energy range 5-7 MeV. ^{18}F is the most appropriate production range; for $^{20}Ne(d,\alpha)$ In nuclear reaction is seen that the energy range 5.5-7.5 MeV. ^{18}F is the most appropriate production range; for $^{16}O(^{3}He,p)$ nuclear reaction is seen that the energy range 6-8 MeV.

Nuclear medicine radioisotopes for production the necessary reaction cross sections were calculated theoretically. The calculated results were compared with the results of experimental data take from literature. As we can understand from the chart, it is seen that the compatible the experimental data taken from literature with the theoretical results calculated using the Talys1.6 program. The findings in this study; it can be evaluated for model the creation of the experimental studies, nuclear reaction cross section calculations, strengthening of data base needed for radioisotope production programs, creation of a new data base, production technology development.

This study contains thesis results of Amasya University Science Faculty master degree.

References

1. V. Çapalı., et al., SDU Journal of Science (E-Journal), 9 (2), 107-112, (2014).
2. A. Koning, S. Hilaire, S. Goriely, TALYS-1.6 A Nuclear Reaction Program, User Manual (NRG, The Netherlands) (2013).
3. EXFOR/CSISRS, Experimental Nuclear Reaction Data File, Brookhaven National Laboratory, National Nuclear Data Center (2009).
4. H.W. Newson., Physical Review,**51**620, (1937).
5. O.D. Brill, L.V. Sumin., Atomnaya Energiya, 7377, (1959).
6. R.J. Jaszczak, R.L. Macklin, J.H. Gibbons, Physical Review, **181** ,1428 (1969).
7. R.W. Michelmann, et al., Nucl. Instrum. Methods in Physics Res., Sect. B, **51**, 1 (1990).
8. I.Ya. Barit, L.E. Kuzmin, A.M. Kazantsev, Journal of Radioanalytical and Nuclear Chemistry, **97**, 97 (1986).
9. M. Guillaume, Nuclear Instrum. and Methods in Physics Res, **136**, 185 (1976).
10. R.L. Hahn, E. Ricci, Physical Review, **146** 650 (1966).
11. J. Fitschen, et al., Applied Radiation and Isotopes, **28**, 781 (1977).

DOI: 10.1051/epjconf/201612801004

Theoretical investigation of cross sections and astrophysical S-factors for the $^{92}Mo(\alpha,n)^{95}Ru$ and $^{94}Mo(\alpha,n)^{97}Ru$ reactions

Abdullah Aydin[1], *Ercan* Yıldız[1], *Ismail Hakki* Sarpün[2]

[1]Kırıkkale University, Physics Department, Kırıkkale, Turkey
[2]Afyon Kocatepe University, Physics Department, Afyonkarahisar, Turkey

Abstract. Molybdenum is commonly applied as a constructive material in different types of nuclear reactors. The cross sections of the $^{92}Mo(\alpha,n)^{95}Ru$ and $^{94}Mo(\alpha,n)^{97}Ru$ reactions have been calculated at 5-20 MeV energy ranges. In theoretical calculations, the TALYS1.6 and NON-SMOKER codes were used. Also the astrophysical S-factors were calculated. Results of our calculations were checked to the experimental data obtained from EXFOR database.

1 Introduction

Molybdenum is significant structural material in fusion reactor technologies and astrophysical applications [1]. Therefore, knowledge of the reaction cross sections and astrophysical S-factors on the Mo isotopes are needed. Cross section measurements for charged-particle capture reactions on nuclei heavier than Fe are important for nucleosynthesis studies [2] and for testing statistical model predictions. Although the descent of the abundances of the elements is generally good understood, significant uncertainties still exist [3]. Therefore, recently (α,n) and (α,γ) reaction cross-sections in low energies are measured by several authors [4-9]. The importance of alpha capture cross sections for different mass regions to test the theoretical models is well known [10]. Experimental measurements of cross section, $\sigma(E)$ at very low energies are mainly not available. The rates can be calculated by extrapolation of the astrophysical S-factor to very low energies or determination of the cross section using theoretical models. The information to be obtained about cross sections or astrophysical S-factors in nuclear astrophysics reactions are the main source of information about the nuclear processes in astrophysics. Various studies using theoretical models have been done to predict the alpha capture cross sections at low energies [3,11].

In this study, we calculated the cross sections and the astrophysical S-factors for $^{92}Mo(\alpha,n)^{95}Ru$ and $^{94}Mo(\alpha,n)^{97}Ru$ reactions up to 20 MeV incident energy. In these theoretical calculations, we used TALYS 1.6 [12] and NON-SMOKER [13] codes. Results of our calculations were checked to the experimental data obtained from EXFOR [14] database.

2 Materials and methods

The charged particle nuclear reaction cross sections is given as

$$\sigma(E)=E^{-1}\exp(-2\pi\eta)S(E) \qquad (1)$$

where E is the center-of-mass energy of the reactants, S(E) is astrophysical factor and $\eta = (Z_1Z_2e^2)/\hbar v$ is the Sommerfeld parameter [15]. Experimental cross section measurements are mainly not available because of the Coulomb barrier. Since the astrophysical S-factor describes the possibility of reaction in low energies, in astrophysical applications, it should be well known for many reactions at low energies ($E \leq$ a few MeV). Also, the astrophysical S-factor is a function of energy with slow variation than $\exp(-2\pi\eta)$ and $\sigma(E)$ [15]. Thus if theoretical astrophysical S-factors are known at low energies, cross sections can be predicted in these energies.

3 Calculations and results

In this study, the reaction cross-sections and the astrophysical S-factors of the $^{92}Mo(\alpha,n)^{95}Ru$ and $^{94}Mo(\alpha,n)^{97}Ru$ reactions were calculated by TALYS 1.6 code and Eq. 2, respectively. The radiative alpha capture reaction cross-sections were calculated in the incident energy range of 5 to 20 MeV, respectively. In these calculations, the TALYS 1.6 and NON-SMOKER codes were used. Also for these reactions, we calculated the astrophysical S-factors which describe the possibility of reaction in low energies. Obtained results were compared to the available experimental data of EXFOR database in Figs. 1–4. It can be seen that there is agreement between the experimental data and calculated results. In $^{92}Mo(\alpha,n)$ and $^{94}Mo(\alpha,n)$ reactions, ^{95}Ru and ^{97}Ru isotopes are produced, respectively. ^{95}Ru isotope ($T_{1/2}$ = 1.64 h) decays

DOI: 10.1051/epjconf/201612801004

to ^{95}Tc ($T_{1/2}$ = 20.0 h) and ^{97}Ru isotope ($T_{1/2}$ = 2.83 days) decays to ^{97m}Tc ($T_{1/2}$ = 91.0 days).

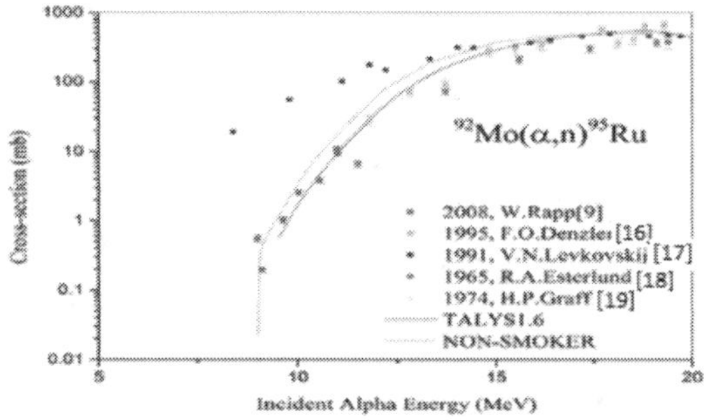

Fig. 1. Comparison of experimental cross sections and theoretical cross sections for $^{92}Mo(\alpha,n)$ reaction

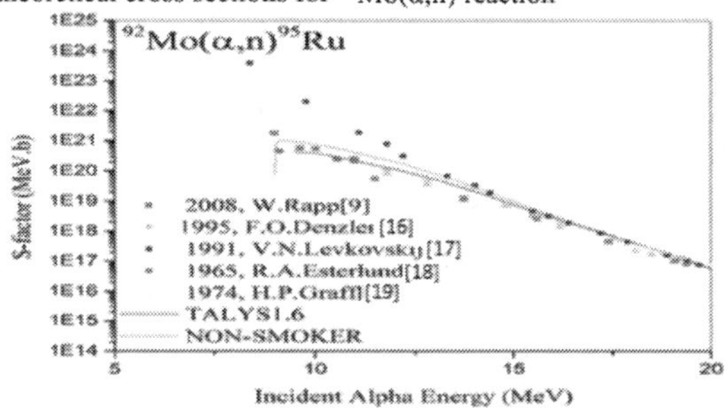

Fig. 2. Comparison of experimental S-factors and theoretical S-factors for $^{92}Mo(\alpha,n)$ reaction

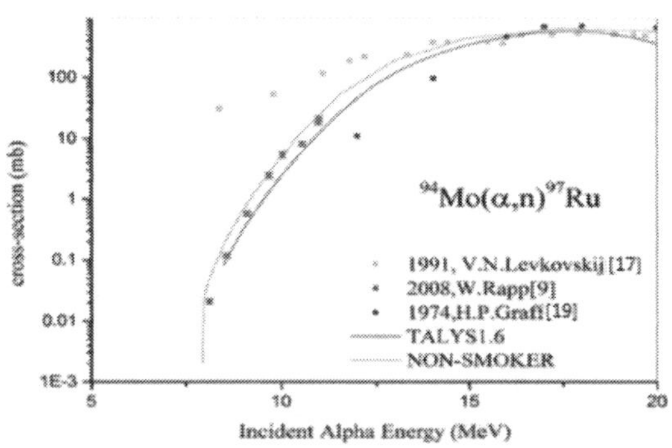

Fig. 3. Comparison of experimental cross sections and theoretical cross sections for $^{94}Mo(\alpha,n)$ reaction

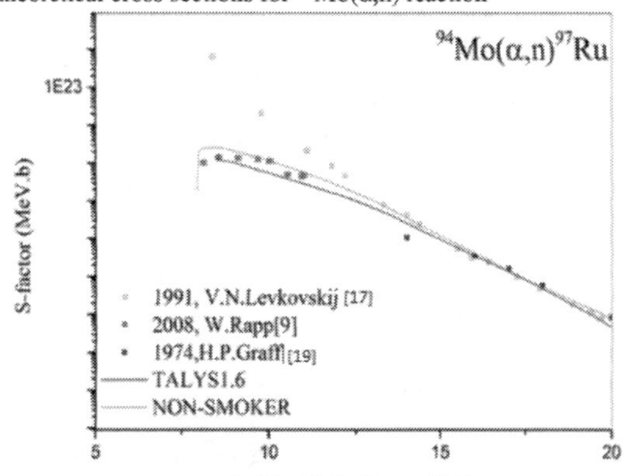

Fig. 4. Comparison of experimental S-factors and theoretical S-factors for $^{94}Mo(\alpha,n)$ reaction

4 Conclusion

The cross-sections and astrophysical S-factors of the $^{92}Mo(\alpha,n)^{95}Ru$ and $^{94}Mo(\alpha,n)^{97}Ru$ reactions have been analyzed at 5-20 MeV. It appears that the agreement between the experimental and evaluated data is reasonable good at the higher energy but poor at the lower energy for these reactions. Therefore, theoretical calculations could be repeated with the new nuclear parameters to obtain the best fit with the experimental data. Also more low-energy experiments are clearly needed for alpha capture reactions in the mass range of nuclei above iron.

References

[1] A. Yu. Smirnov et al., Physics Procedia 72, 126 – 131, (2015).
[2] A. Simon et al., Phys. Rev. C 92, 025806, (2015).
[3] N. Ozkan et al., Nucl. Phys. A 710, 469, (2002).
[4] A. Simon et al., Phys. Rev. C **92**, 025806, (2015).
[5] M.S. Basunia et al., Phys. Rev. C **71**, 035801, (2005).
[6] G.G. Kiss et al., Phys. Lett. B **735**, 40, (2014).
[7] Gy. Gyürky et al., J. Phys. G **37**, 115201, (2010).
[8] A. Sauerwein et al., Phys. Rev. C**84**, 045808, (2011).
[9] W. Rapp et al., Phys. Rev. C **78**, 025804, (2008).
[10] W. Rapp et al., Phys. Rev. C **66**, 015803, (2002).
[11] E. Yildiz et al., EPJ Web of Conferences **100**, 01010, (2015).
[12] http://www.talys.eu
[13] http://nucastro.org/nonsmoker.html
[14] http:// www-nds.iaea.org/exfor/exfor.htm
[15] D.G. Yakovlev et al., Phys. Rev. C82, 044609, (2010).
[16] F.O. Denzler et al., Radiochimica Acta Vol.68, p.13, (1995).
[17] V.N. Levkovskij, Activation cross section nuclides of average masses (A=40-100)by protons and alpha-particles with average energies (E=10-50 MeV), Moscow, (1991).
[18] R.A. Esterlund et al.,Nuclear Physics Vol.69, Issue.2, p.401, (1965).
[19] H.P. Graf et al., Journal of Inorganic and Nuclear Chemistry Vol.36, p.3647, (1974).

EPJ Web of Conferences **128**, 01005 (2016) DOI: 10.1051/epjconf/201612801005
TESNAT 2016

Calculation of (p,γ) and (p,α) nuclear reaction cross sections in stars up to 10 MeV

Eyyup Tel[1], *Muhittin* Sahan[1], *Ismail Hakki* Sarpun[2], *Sureyya Gulistan* Okur[1]

[1]Osmaniye Korkut Ata University, Faculty of Arts and Science, Department of Physics, Osmaniye, Turkey
[2] Afyon Kocatepe University, Faculty of Arts and Sciences, Department of Physics, Afyonkarahisar, Turkey

Abstract. In Knowledge of the proton-proton (p-p) chain and CNO cycle are required for the evolution of main sequence stars during the early formation of the universe. In this study, we summarized the excitation functions of (p,γ) and (p,α) reactions for $^{7}Be(p,\gamma)^{8}B$, $^{12}C(p,\gamma)^{13}N$, $^{13}C(p,\gamma)^{14}N$, $^{14}N(p,\gamma)^{15}O$ and $^{15}N(p,\alpha)^{15}O$ in p-p chain and CNO cycle using EMPIRE and TALYS computer up to 10 MeV. The calculated data on nuclear fusion cross sections in hydrogen-burning stars were compared with theoretical TENDL-2014 and ENDF/B-VII data from EXFOR. The calculation results show closed agreement between the calculations and the data from literature.

1 Introduction

Animals, plants, soil, air, planets, stars like everything formed by the basic blocks of matter called atoms. After the big bang, approximately 13.7 billion years ago, a large part of atoms formed in the early formation of the universe and came unchanged until the present day. Since the universe includes everything from subatomic particles to galaxies, from planets to stars, it is needed to be explored in detailed.

All stars are born from collapsing clouds of gas and dust the interstellar medium (ISM). During their lives, they deposit energy into the ISM in the form of electromagnetic radiation and stellar winds. When they die, they return some of their matter and energy back into the ISM. The life time of stars range from a few million years for the most massive to trillions of years for the least massive [1-3].

The amount of H and He elements in the universe are approximately 73% and 25% of the mass, respectively. After H and He, C, N and O elements are the most abundant elements in the Universe. Energy production in massive stars is governed by the CNO cycles throughout most of their lifetimes [4]. Hydrogen and helium elements having low-mass were produced in the hot and dense conditions of the birth of the universe itself. The life of stars is described in terms of nuclear reactions.

The Hydrogen burning mechanism is essentially the nuclear fusion of four protons into one ^{4}He nucleus. There are two reaction chains that can convert hydrogen to helium, namely the proton–proton (or pp chain) chain, also called pp reaction, and the CNO cycle. The pp chain is more important in stars the mass of the Sun or less and stars of similar mass The CNO cycle by which stars convert hydrogen to helium is the dominant source of energy in more massive stars [5].

The p-p fusion is the nuclear fusion process which fuels the Sun and other stars which have core temperatures less than 15×10^{6} K and a reaction cycle at energy of 25 MeV. The p-p chains occur under lower temperature than the carbon-nitrogen-oxygen cycle. The pp chain in stars like our Sun divides into three main branches, called as the ppI, ppII and ppIII chains. These are given Eq. 1-3. According to Eq. 1, in the *proton-proton* fusion reaction inside stars like the sun, the first step in ppI process is the collision of two protons (p or ^{1}H) to make an atom of heavy hydrogen (^{2}H). The second step containing the weak interaction involves the collision of a ^{2}H nuclei with a proton to make a nucleus of ^{3}He, and then gamma ray is emitted having very high energy. In the final third step, a normal ^{4}He nucleus are created with combining two ^{3}He nuclei and two extra protons (2^{1}H) are released to start the whole process again. Inside the Sun, about 655×10^{6} tons of H are converted into 650×10^{6} tons of ^{4}He every second. In stars heavier than about 2 solar masses, in which the core temperature is more than about 18×10^{6} K, the dominant process in which energy is produced by the fusion of hydrogen into helium is a different reaction chain known as the carbon-nitrogen cycle. The former case is the last reaction of the ppI chain, whereas the latter reaction leads into either the ppII or the ppIII chain, as given Eq. 2-3.

The second process of stable hydrogen burning reaction in stars is the CNO cycle, dominates from 20×10^{6}K to 130×10^{6}K. This cycle for star more massive than the Sun converts hydrogen to helium according to the following sequence of reactions in Eq. 4. The CNO cycle occurs from the following steps. After ^{12}C element captures a proton, ^{13}N is produced and gamma-ray is emitted. ^{13}N is unstable and decays to the ^{13}C with a half-life of about 10 minutes. The ^{13}C isotope captures a proton and a gamma-ray is emitted to become ^{14}N isotope. ^{14}N isotope captures a proton and a gamma-ray is emitted to become ^{15}O. ^{15}O isotope undergoes a β^+ decay to become ^{15}N. Finally, ^{15}N isotope captures a proton and then it emits a ^{4}He to close the cycle and return to ^{12}C.

ppI chain:

$$
\begin{aligned}
&{}^1_1H+{}^1_1H\rightarrow{}^2_1H+e^++\nu_e\\
&{}^2_1H+{}^1_1H\rightarrow{}^3_2He+\gamma \qquad (1)\\
&{}^3_2He+{}^3_2He\rightarrow{}^4_2He+2{}^1_1H
\end{aligned}
$$

ppII chain:

$$
\begin{aligned}
&{}^3_2He+{}^4_2He\rightarrow{}^7_4Be+\gamma\\
&{}^7_4Be+e^-\rightarrow{}^7_3Li+\nu_e \qquad (2)\\
&{}^7_3Li+{}^1_1H\rightarrow2{}^4_2He
\end{aligned}
$$

ppIII chain:

$$
\begin{aligned}
&{}^3_2He+{}^4_2He\rightarrow{}^7_4Be+\gamma\\
&{}^7_4Be+{}^1_1H\rightarrow{}^8_5B+\gamma \qquad (3)\\
&{}^8_5Be\rightarrow{}^8_4Be+e^++\nu_e\\
&{}^8_4Be\rightarrow2{}^4_2He
\end{aligned}
$$

CNO Cycle

$$
\begin{aligned}
&{}^{12}_6C+{}^1_1H\rightarrow{}^{13}_7N+\gamma\\
&{}^{13}_7N\rightarrow{}^{13}_6C+e^++\nu_e\\
&{}^{13}_6C+{}^1_1H\rightarrow{}^{14}_7N+\gamma \qquad (4)\\
&{}^{14}_7N+{}^1_1H\rightarrow{}^{15}_6O+\gamma\\
&{}^{15}_6O\rightarrow{}^{15}_7N+e^++\nu_e\\
&{}^{15}_7N+{}^1_1H\rightarrow{}^{12}_6C+{}^4_2He
\end{aligned}
$$

TALYS computer code system in the from 1 keV to 1 GeV energy range [6] and EMPIRE computer code over from keV up to few hundred MeV energy range [7] have been widely used in order to calculate nuclear reactions. In this study, in order to calculated the excitation functions of (p,γ) and (p,α) cross section for some reactions ^{15}O in pp chain and CNO cycle, we used EMPIRE and TALYS computer up to 10 MeV. Results were compared with theoretical TENDL-2014 and ENDF/B-VII data from EXFOR [8].

2 Results and Discussions

In this study, in order to calculated the excitation functions of (p,γ) and (p,α) cross section for some reactions such as ^{7}Be(p,γ), ^{12}C(p,γ), ^{13}C(p,γ), ^{15}N(p,a), ^{14}N(p,γ) in pp chain and CNO cycle, we used two computer code systems called as TALYS-1.6 [6], EMPIRE-3.2.2-Malta [7] up to 10 MeV. Results were compared with theoretical TENDL-2014 and ENDF/B-VII data from EXFOR from EXFOR library [8].

Our results from calculations are given Figure 1-5. Theoretical calculated cross sections of ^{7}Be(p,γ) reaction in ppIII chain between 0-10 MeV is shown in the Fig. 1. This reaction's cross section data obtained from EMPIRE and ENDF that shows to us increasing values with 0.75-5 MeV energy ranges. However, in spite of increasing these values, reaction cross section value from TALYS was a pick at about 5 MeV energy. At the same time, the data from EMPIRE and TALYS computer code, and also ENDF systems have showed increasing values in the 6.5 MeV with 10 MeV energy range.

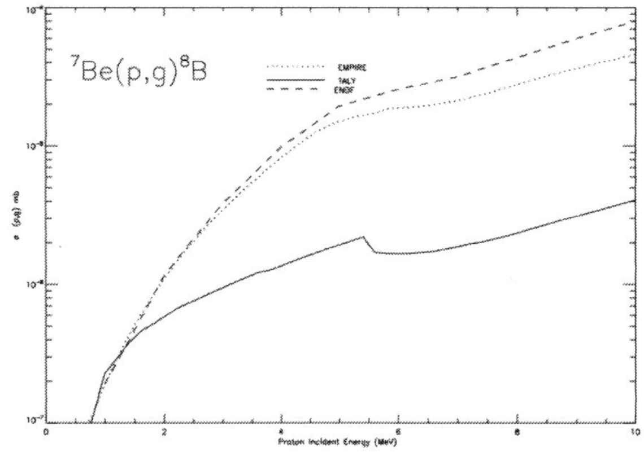

Fig. 1. Theoretical cross sections calculation of ^{7}Be(p,γ) reaction up to 10 MeV.

Reaction cross section of ^{12}C(p,g) accruing in the CNO cycle is given in Fig. 2. As seen from Fig. 2, ENDF and TALYS theoretical data are in harmony between 2 and 5 MeV. The data from EMPIRE also joins at this harmony between 3 and 5 MeV. Only one experimental data of Burtebaev [9] could be found from EXFOR in literature at approximately 4.27 MeV. This data is closed agreement with ENDF at around 4 MeV.

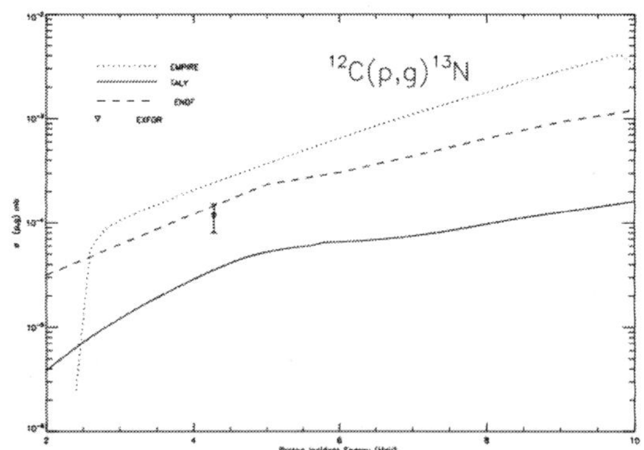

Fig. 2. Theoretical cross sections calculation of ^{12}C(p,γ) reaction up to 10 MeV.

Reaction cross section of ^{12}C(p,γ) in the CNO cycle is given in Fig. 3. When ^{13}C(p,g) reaction was analyzed, while it can be seen disharmony between EMPIRE, TALYS, and ENDF computer data up to 2.5 MeV, there is a good harmony over 2.5 MeV. Also, while the line from EMPIRE decreases up to 2.5 MeV, the line from

TALYS increases. There is no experimental data for $^{12}C(p,g)$ reaction from literature. The fourth reaction cross section in the CNO cycle is $^{14}N(p,\gamma)$ given in Fig. 4. It can be seen from Figure 4, EMPIRE line decreases up to about 5 MeV while TALYS and ENDF lines increase. All cross sections data are approximately the same values between 5-10 MeV. The fifth and final reaction cross section calculation in the CNO cycle is $^{15}N(p,\alpha)$ as given in Fig. 5. For three lines, it can be seen that there is a good relation from 0 to 10 MeV. All cross section data have a rising trend up to about 5 MeV but having a stable and constant values after 5 MeV.

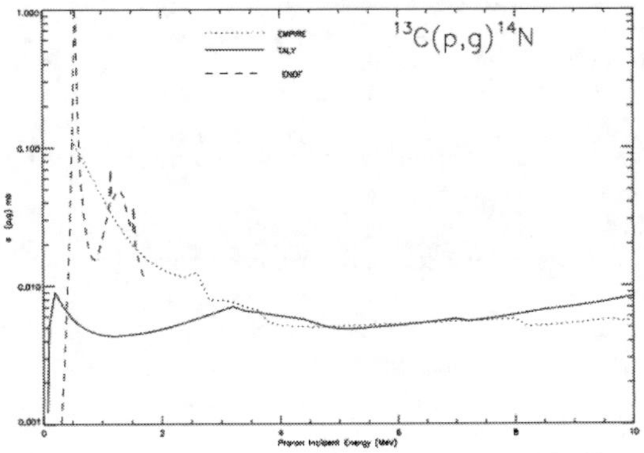

Fig. 3. Theoretical cross sections calculation of $^{13}C(p,\gamma)$ reaction up to 10 MeV.

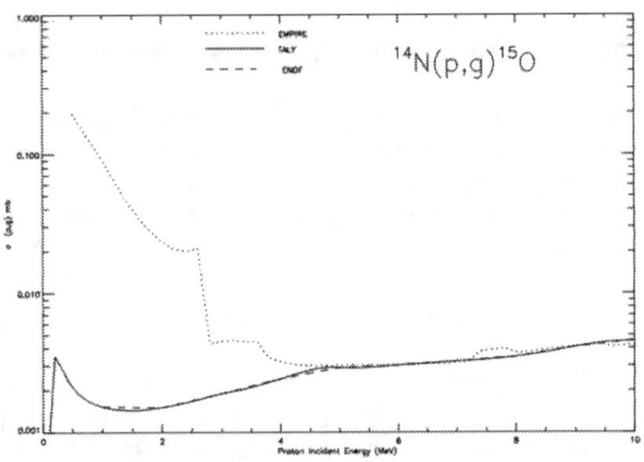

Fig. 4. Theoretical cross sections calculation of $^{14}N(p,\gamma)$ reaction up to 10 MeV.

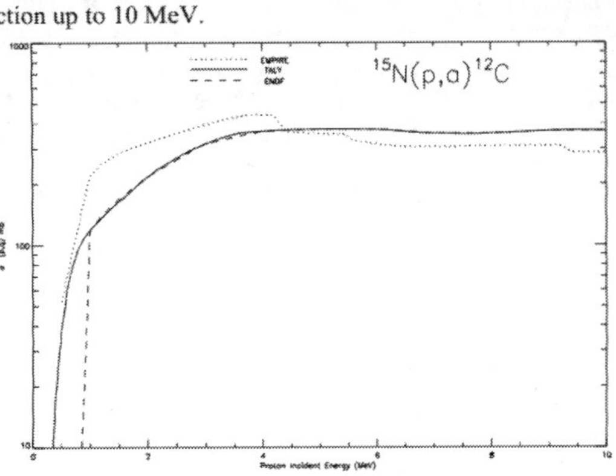

Fig. 5. Theoretical cross sections calculation of $^{15}N(p,\alpha)$ reaction up to 10 MeV.

3 Conclusion

We know that it is necessary hydrogen burning to be able to start the nuclear fusion of four protons into one ^4He nucleus in stars. In this work, (p,γ) and (p,γ) cross section reactions for ^7Be, ^{13}C, ^{15}N, and ^{14}N nuclei in pp chain and CNO cycle in stars were calculated in the energy range from 0 to 10 MeV using TALYS 1.6 [7] and EMPIRE-3.2.2 [8] computer codes. Since these reactions in pp chain and CNO cycle occur in low energies, it is difficult to calculate cross sections in these nuclei. We were able to calculated only five limited elements in in pp chain and CNO cycle. We suggest that we can achieve better results when added resonance effects or resonance parameters in nuclear code calculating programs.

This work is supported by the Coordinator ship of Osmaniye Korkut Ata University's Scientific Research Office (BAP) Contract No. *OKÜBAP-2015-PT3-011*. The authors would like to thank to Prof.Dr. Arjan Koning for supporting TALYS program code.

References

1. K. M. Ferriere, Rev. Mod. Phys., 73, 1031 (2001)
2. Reynolds R. J., Science, 277, 1446 (1997)
3. M.Sahan, et al., Chin. J. Astron. & Astrophys. 5, 211 (2005).
4. N. Przybilla, et al., Astronomy and Astrophysics 517, A38 (2010).
5. M.C. Santi, Evolution of stars and stellar populations, Wiley (2005).
6. A. Koning, S. Hilaire, S. Goriely, TALYS-1.6-A Nuclear Reaction Program, User Manual, 1st edition NRG, The Netherlands, (2013).
7. M. Herman et al., EMPIRE-3.2 Malta code, User's Manual. (http://www.nndc.bnl.gov/empire (2013).
8. Experimental Nuclear Reaction Data (2015), <http://www.nndc.bnl.gov/exfor/exfor.htm>
9. N.Burtebaev, et al., Phys. Rev. C 78, 035802 – Published 11 September (2008)

DOI: 10.1051/epjconf/201612802001

Preliminary results of Sr/Ca ratio study of teeth samples by photoactivation analysis

Yusuf Kavun[2,3], *Ismail* Boztosun[1,2], *Ismail* Maraş[3], *Haris* Đapo[1,2], and *Christian* Segebade[4]

[1] Department of Physics, Akdeniz University, TR-07058, Antalya, Turkey
[2] Nuclear Sciences Application and Research Centre, Akdeniz University, TR-07058, Antalya, Turkey
[3] Department of Physics, Celal Bayar University, TR-45030, Manisa, Turkey
[4] Idaho Accelerator Centre, Idaho State University, 921 S. 8th Ave., Pocatello, ID 83209, United States

Abstract. In this paper, we have performed Photoactivation Analysis (PAA), a non-destructive method, which is used to determine elemental concentration of any sample. This paper presents the first use of this method in medical sciences in Turkey. The method was applied to the determination of Sr/Ca ratios in teeth. The collected teeth samples and standards (SrO and $CaCO_3$) have been irradiated for a fixed time interval with high energy photons. The photons were generated by a clinical linear accelerator (cLINAC). The photon end-point energy was 18 MeV. The energy and the time interval were sufficient to achieve good activation. Afterward, the samples and standards have been analysed with gamma spectroscopic analysis by using an HPGe detector system. By analysing many samples, a database of Sr/Ca ratios will be created at Nuclear Research and Application Center (NUBA). In this paper we present a small subset of the already analysed data as an example of our capabilities and goal. We hope to set an example for future studies.

1. Introduction

The interest in the lifestyles of ancient societies is increasing day by day. Among these studies, determination of nutrition models, the discovery of migration patterns, achieve the population structure, constructed of community way of life, the relationship between diseases and lifestyle, determination of individual growth and development through identification of environmental compliance strategy holds an important role. Especially, teeth provide more information to investigate the effects on humans of environmental factors, disease and nutrition. In order to avoid biological, environmental, and archaeological specimens' damage to the preservation of basic and trace elements and values that are available for non-destructive analysis is very important. There are alkaline earth metals present in the structure of teeth like Strontium (Sr), Calcium (Ca), Barium (Ba) and Magnesium (Mg) [1, 14]. Of these alkali metals, Sr/Ca ratio is very useful parameter. Sr/Ca ratio, determination of the effect of environmental food science and nutrition, breastfeeding offers focused on the biological importance of human health such as diagnosis and treatment process. The ratios of strontium (Sr) and calcium (Ca) elements which are found abundantly in teeth will have benefits in biological, archaeological and anthropological studies. This ratio having chemically similar, often used to understand the influence of the environment [2,13] such as previous life of dietary information [4,11,12], cutting applications milk [5,9], food and disease event. Besides, this study is expected to provide an insight on health issues like eating habits -therefore welfare levels- of people within different time frames, effects of breastfeeding, environmental influence on disease and nutritional status, determination of age-race, evaluation of teeth quality, and evaluation of treatment success of teeth recovery. Sr concentration was used to examine the dietary in prehistoric human populations. Also it is used to determine to rate of consumed meat and plants [6-8]. Method here, in order to support the calcium, mammalian digestive system provides discrimination to strontium [10]. In addition, Mammalian kidneys filter out the strontium faster than calcium from the body. Therefore, individuals who consume large amounts of meat is supposed to be have amounts of strontium in skeletons [8]. This discrimination against strontium (or calcium bio-treatment) has been proven in the High Sierras where full of carnivores, herbivores and plant in the natural food chain in a discrete ecosystem of the alpine foothills [3].

2. Material and Method

Teeth samples were collected from the patients of Akdeniz University Faculty of Dentistry. The collected samples and prepared standards (SrO and CaCO3) were irradiated for a specific period of time with a clinical linear accelerator (cLINAC) of 18 MeV bremsstrahlung energy to achieve photo-activation. The prepared calibration material was controlled with universally reference material (URM, table 1). Afterwards, the samples and standards are analyzed with gamma

DOI: 10.1051/epjconf/201612802001

spectroscopic analysis method using an HPGe detector system and a database is formed.

The teeth samples and reference materials are prepared in the same geometry and then placed 56 cm away from the irradiation zone to perform irradiation (Fig. 1). The electrons obtained from electron gun, hit the tungsten target resulting in bremsstrahlung photons and are directed with primary collimators towards the flattening filter (Fig. 2). Wherein, the photon flux along the dotted line has the highest value in Fig. 2. Therefore, by using of a smoothing filter with a triangular geometry a spatial homogeneous photon flux distribution is achieved. After passing the ionization chamber used in dosimetry, photons eventually (Y and X aperture are fully open) form a 40x40 cm^2 square at 100 cm distance. The width of the field can be expanded or reduced by the diaphragm Y and X.

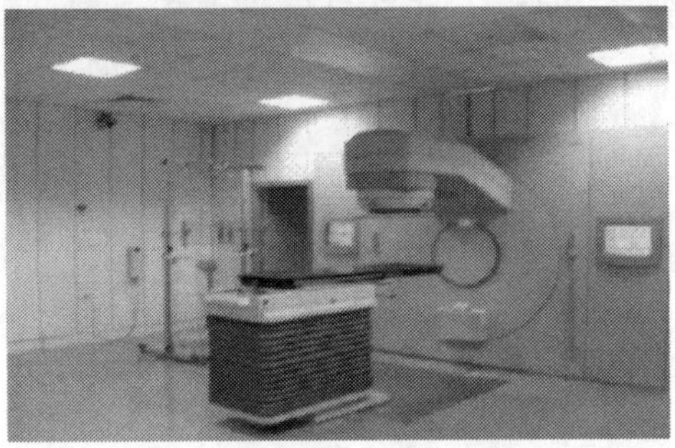

Fig. 1: Philips SLI-25 clinical linear electron accelerator of Elekta TM Synergy TM.

By making the necessary settings from the control room, it is possible to ensured same exposure to sample and reference. Over a course of two hours, samples are subjected to bremsstrahlung photons with sufficient energy (usually between 18 and 30 MeV) to induce photonuclear reactions. Both samples are assumed to be subjected to the same amount of photon flux during this process. A second assumption is that the isotopes in both of the analyzed samples are found in their natural abundance. Based on these assumptions and the model of activation analysis, the relative concentrations of the elements in the examined samples is calculated from equations (1) and (2) [9]:

$$c_s = F_{corr} \cdot \frac{c_{cal} \cdot m_{cal} \cdot P_s}{m_s \cdot P_{cal}} \cdot \frac{e^{-\lambda\left(t_{d_{cal}} - t_{d_s}\right)} - e^{-\lambda\left(t_{d_{cal}} - t_{d_s} \; t_{c_{cal}}\right)}}{1 - e^{-\lambda \cdot t_{c_s}}} \quad (1)$$

$$F_{corr} = \frac{(m_{Mon1Sm} + m_{Mon2Sm}) \cdot \left(\frac{I_{Mon1Cal} \cdot \lambda_{Mon}}{e^{-\lambda_{Mon} \cdot t_{dMon1Cal}} \cdot (1 - e^{-\lambda_{Mon} \cdot t_{cMon1cal}})} + \frac{I_{Mon2Cal} \cdot \lambda_{Mon}}{e^{-\lambda_{Mon} \cdot t_{dMon2Cal}} \cdot (1 - e^{-\lambda_{Mon} \cdot t_{cMon2Cal}})} \right)}{(m_{Mon1Cal} + m_{Mon2Cal}) \cdot \left(\frac{I_{Mon1Sm} \cdot \lambda_{Mon}}{e^{-\lambda_{Mon} \cdot t_{dMon1Sm}} \cdot (1 - e^{-\lambda_{Mon} \cdot t_{cMon1Sm}})} + \frac{I_{Mon2Sm} \cdot \lambda_{Mon}}{e^{-\lambda_{Mon} \cdot t_{dMon2Sm}} \cdot (1 - e^{-\lambda_{Mon} \cdot t_{cMon2Sm}})} \right)} \quad (2)$$

where c_{cal} is the concentration of the element under study in the calibration material, m_{cal}, m_s are the mass of calibration and sample, P_s, P_{cal} are the net peak integral of the analyses to be determined in the sample and calibration material, λ is the decay constant of the radionuclides (^{47}Ca, ^{87m}Sr) produced during photon activation, t_d, t_c are decay and counting period of the sample/calibration material, M_{mon} is the masses of the bremsstrahlung flux monitors, I_{mon} is the annihilation net peak integrals of the monitor spectra, λ_{mon} is the decay constant of the monitor radionuclide (^{64}Cu) produced during photon activation, t_{dMon}, t_{cMon} are decay and counting period of the monitors. The equation for uncertainty calculation is (3):

$$U_c(Elem.)\% = \sqrt{K_1^2 + K_2^2 + K_3^2 + \ldots + K_n^2} \quad (3)$$

K is values of the uncertainties of the Ca and Sr results.

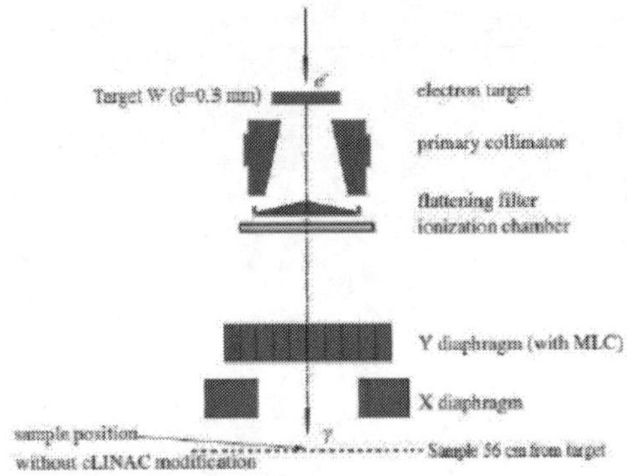

Fig. 2: Schematic view of the cLINAC

Table 1. Comparison of Universally Reference material (URM) and prepared calibration material by Schmitt B. F., Segebade C. and Fusban H.U.

Element	^{87}Sr	^{48}Ca
Isotope	^{87m}Sr	^{47}Ca
keV	388	1297
Peak Sm.	266161.5	2472.83
Conc. Cal.	89258.21	78818.38
Conc. Sm.	30999 µg/g	68.85%
Lit. Value	778 ±38 µg/g	10.16% ±0.18%

3. Results and Discussion

In this study age dependence of the Sr/Ca ratio in 10 human teeth were examined. Photo-activation analysis was used for the detection of trace elements and amount of Sr/Ca ratio. Changes with respect to age and the effect of diet and environmental factors were investigated. This study is an example of first use of photo-activation analysis in our country for studies relating to health science. The value of Spearman's rho correlation (aka fit correlation coefficient) is -0.426 that calculated by using SPSS (Table 2). This value indicates a high negative correlation between age and Sr/Ca ratio. Sr/Ca ratio decreases with increasing age in the teeth. These relations are too high to be ignored. Because of the age factor can be explained only 9% of Sr/Ca rate. Calculated concentrations of Ca and Sr, and derived Sr/Ca ratios, in human teeth is shown in Table 3.

The remaining 91% factors are influenced by different factors such as climate, nutritional factors, genetic factor, gender and diet. While the age increase, the Sr/Ca ratio of teeth is decreasing 1/0.0026.

Table 2: The Value of Spearman's rho Correlation is calculated by SPSS

Correlations

			SrCa	Age
Spearman's rho	Sr/Ca	Correlation Coefficient	1.000	-.426
		Sig. (2-tailed)	.	.220
		N	10	10
	Age	Correlation Coefficient	-.426	1.000
		Sig. (2-tailed)	.220	.
		N	10	10

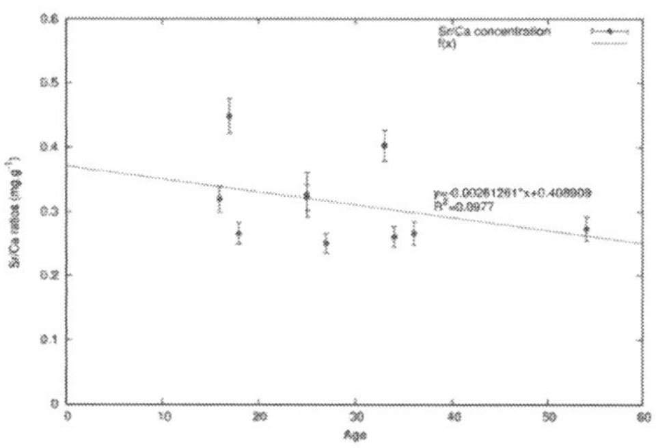

Fig. 3. The change of Sr/Ca ratios in teeth depending on age.

Table 3. Concentrations of Ca and Sr, and derived Sr/Ca ratios, in human teeth.

Sample No	Age	Elemental Concentration Ca (%)	Elemental Concentration Sr($\mu g.g^{-1}$)	Sr/Ca($mg.g^{-1}$)
1	34	41(4)	106(5)	0.261(0.016)
4	25	43(5)	140(6)	0.322(0.021)
7	17	40(4)	179(8)	0.448(0.027)
8	33	40(4)	161(7)	0.403(0.025)
9	18	51(4)	136(7)	0.266(0.017)
10	54	37(4)	102(6)	0.274(0.019)
11	27	40(4)	99(5)	0.251(0.016)
13	36	35(4)	94(5)	0.266(0.018)
14	25	41(4)	220(11)	0.53(0.04)
17	16	40(4)	128(6)	0.320(0.020)
Mean ±SD				0.334±0.021

References

1. J.H. Chao, et al., Applied Radiation and Isotopes, **67**, 1121–1126, (2009)
2. C.J. Brown, et al., Archives Of Oral Biology, **49,** 705–717, (2004)
3. R.W. Elias, Y. Hirao, C.C. Patterson, Geochimica et Cosmochimica Acta, **46**, 2561-2580, (1982)
4. M. Sponheimer, et al., J. Of Human Evolution, **48**, 147–156, (2005)
5. A. Sillen, P. Smith, J. of Archaeological Science, **11**, 237–245,(1984)
6. M.J. Schoeninger, American J. of Physical Anthropology, **51**, 295-310,(1979)
7. T.D. Price, M. Connor, J.D. Parsen, J. of Archaeological Science, 12, 419-442,(1985)
8. J.C. Sealy, A. Sillen, J. of Archaeological Science, **15**, 425-438,(1988)
9. Z.J. Sun, et al., AIP Conf. Proc. 1336, 473.2011
10. K. Kostial, N. Gruden, A. Durakovic, Calcified Tissue Research, **4**, 13-19, (1969)
11. K. Ozdemir, Y.S. Erdal, Çocuk Sağlığı ve Hastalıkları Dergisi, **52**, 128-140, (2009)
12. H. Toots, M.R. Voorhies, Science, **149**, 854–855, (1965)
13. E. Webb, et al., Micro chemical Journal, **81**, 201–208, (2005)
14. S. Mays, J. of Archaeological Science, **30**, 731–741, (2011)

Quantum chemical calculations and spectroscopic measurements of spectroscopic and thermodynamic properties of given uranyl complexes in aqueous solutions with possible environmental and industrial applications

Jakub Višňák[1,2,3], *Lukáš* Sobek[1,3]

[1]*Dep. of Nuclear Chemistry, FNSPE, Czech Technical University, Břehová 7, 115 19 Prague 1, Czech Rep.*
[2]*Dep. of Chemical Physics and OpticsCharles University, Ke Karlovu 3, 121 16 Prague 2, Czech Rep.*
[3]*J. Heyrovský Institute of Physical Chemistry, Dolejškova 2155/3, 182 23 Prague 8, Czech Rep.*

Abstract. A brief introduction into computational methodology and preliminary results for spectroscopic (excitation energies, vibrational frequencies in ground and excited electronic states) and thermodynamic (stability constants, standard enthalpies and entropies of complexation reactions) properties of some 1:1, 1:2 and 1:3 uranyl sulphato- and selenato- complexes in aqueos solutions will be given. The relativistic effects are included via Effective Core Potential (ECP), electron correlation via (TD)DFT/B3LYP (dispersion interaction corrected) and solvation is described via explicit inclusion of one hydration sphere beyond the coordinated water molecules. We acknowledge limits of this approximate description – more accurate calculations (ranging from semi-phenomenological two-component spin-orbit coupling up to four-component Dirac-Coulomb-Breit hamiltonian) and Molecular Dynamics simulations are in preparation. The computational results are compared with the experimental results from Time-resolved Laser-induced Fluorescence Spectroscopy (TRLFS) and UV-VIS spectroscopic studies (including our original experimental research on this topic). In case of the TRLFS and UV-VIS speciation studies, the problem of complex solution spectra decomposition into individual components is ill-conditioned and hints from theoretical chemistry could be very important. Qualitative agreement between our quantum chemical calculations of the spectroscopic properties and experimental data was achieved. Possible applications for geochemical modelling (e.g. safety studies of nuclear waste repositories, modelling of a future mining site) and analytical chemical studies (including natural samples) are discussed.

1 Motivation

Uranium and, in particular, soluable uranyl complex compounds are of great importance for both nuclear energetics and environmental chemistry.

Thermodynamic properties of several uranyl complexes (stability constants, $\beta°$, standard complexation enthalpies $\Delta H°$, entropies $\Delta S°$ and the interaction parameters $\varepsilon(i,j)$ (from the Specific Ion Interaction Theory [70-72])) are important input parameters for geochemical modelling studies of various systems ranging from homogenous nuclear reactors to radioactive waste repositories (RWR) or uranium mining sites.

Spectroscopic properties are important for chemical analysis/detection of the previously mentioned complexes. This include use in natural waters chemical analysis (near the mining sites, RWRs or in environment in general). We were concerned in UV-VIS spectrophotometry and Time-Resolved Laser-Induced Fluorescence Spectroscopy (TRLFS) [1].

The above mentioned thermodynamic stability constants $\beta°$ are assigned to chemical reactions

$$UO_2\left(H_2O\right)_5^{2+} + m\,XO_p^{2-} \quad UO_2\left(XO_p\right)_m\left(H_2O\right)_{5-f}^{2-2m} + f\,H_2O, \tag{1}$$

where X = S, Se, C for sulfates, selenates (p=4) and carbonates (p=3) respectively. Sulfates and carbonates are the most common forms of uranium in the early phases of acid and basic leaching in uranium exploitation [2,3]. Selenates are of importance in RWR studies since the ^{79}Se isotope is thermal fission product of critical importance (in particular because of its mobility and radiological and chemical toxicity [4]). The comparison between uranium sulfates, selenates (and tellurates) is important for pure chemistry as well, as an example of homological series of ligands in coordination chemistry.

The usual experimental methodic for $\beta°$ derivation is a speciation study [5-11]. The study consists of preparation of series of solutions all contaning fixed total concentration of uranium c_U, pH and ionic strength I, but differing in the total concentration of ligand (e.g. added in the form of its respective sodium salt – Na_2SO_4, Na_2SeO_4,

 DOI: 10.1051/epjconf/201612802002

Na_2CO_3) and spectral measurement on each sample in the series (temperature *T* is fixed, the number of solutions in the series range from 7 to several tens). Each solution will contain equilibrium mixture of UO_2^{2+}, UO_2XO_p, $UO_2(XO_p)_2^{2-}$, $UO_2(XO_p)_3^{4-}$. The task is now to determine both individual complex spectra and the stability constants (determining the ratios of complex concentrations) at the same time from the series of above mentioned solutions spectra. The task is possible under the assumption that individual component spectra are independent of the solution composition and the solution spectra are therefore just linear combinations of the unknown individual component spectra. The mathematical solution to this problem is long time known (studied in Chemometrics) and we can address interested reader to literature [12-15]. Our approach was based on the (Parallel) Factor Analysis methods [16-18] (which exploits the Singular Value Decomposition [19]).

Because the stability constants of 1:1, 1:2 and 1:3 complexes are very similar and rather small, solutions where only two complex species are nonneglibly present are rather rare. Unfortunately, also the individual component spectra (both the UV-Vis absorption and TRLFS) are very similar to each other [5,8,13], so the numerical analysis of the speciation study spectra is an ill-conditioned problem (i.e. very small variation/error in the experimental data propagate as a large error of the both individual component spectra and the stability constants).

To determine the true value of β^0 (defined by products/reactants concentration ratio extrapolated to zero ionic strength) and $\varepsilon(i,j)$ we would need the above mentioned study to be done for a *serie of series* (the ionic strength is varied, e.g., 0.3, 1, 2, 3, 7, 10 mol·kg^{-1}). Roughly speaking, we can derive β^0 and $\Delta\varepsilon$ values, specific for each complex from SIT-fit of the dependence of β on I (where β is defined by the concentration ratio for non-zero ionic strength). Similarly, the serie of series (when the temperature is varied) study can uncover the values of ΔH^0 and ΔS^0 through Van't Hoff fit (39) [20].

Our idea was to use the theoretical, *ab initio*, predictions of both spectroscopic (parameters of the individual component spectra) and thermodynamic properties as a hint (or constrain) in the decomposition of experimental solution spectra. Later, we have found that a similar approach was used for uranyl hydroxo- complexes by Drobot et al [106]. We also look for quantum chemistry to interpret the experimental results. The quantum chemical modelling of room temperature water solved uranyl complexes deals with – relativistic quantum effects (due to uranium atom, [21]), relatively strong electron correlation and needs for precise description of the solvation (hydration) and molecular dynamics. We were not able to produce numerical accurate (quantative) agreement between theory and experiment, but we would like to present some methodology which allows further systematic improvement to achieve that later (substantiated by qualitative agreement of some spectroscopic quantities achieved yet).

2 Theoretical background

2.1 Uranyl compounds spectra

The studied regions in UV-vis and in emission-wavelength domain of the TRLFS spectra were 450 nm-600 nm, 380 nm-500 nm respectively (or 16.5-22 and 20-26 thousand cm^{-1}). In this regions we can think about a single electronic transition corresponding for the whole spectrum[a]. This transition is vibrationally resolved with peaks corresponding to different vibrational quantum numbers of the initial and final states in question. The vibrational "substates" of the excited electronic state will be labeled by primed numbers (i.e. 0', 1', 2', ...), while the vibrational "substates" of the ground electronic state by non-primed numbers (i.e. 0, 1, 2, ...). The simplest model suppose that electronic excitations in all studied uranyl complexes are located on the central uranyl (UO_2^{2+}) group and since the HOMO(σ_u)→LUMO(δ_u,ϕ_u) electronic excitation corresponds to the transfer of electronic density from bonding σ_u molecular orbital (localized on the O-U-O uranyl axis) to non-bonding orbitals localized mostly on uranium (δ_u,ϕ_u), we can think that only symmetric vibrational mode of uranyl group force constant will be significantly affected by the electronic excitation. And, therefore, only this mode will "vibrationally resolve" the spectrum. For the simplest model, we calculated the symmetric dissociation curve for the "bare" uranyl group (UO_2^{2+}) in vacuum (Fig.1), the corresponding experimental emission/luminescence spectrum of hydrated uranyl ($[UO_2(H_2O)_5]^{2+}$) is presented in Fig. 2. The molecular orbitals discussed above have been calculated for the equilibrium geometry within the same methodics as the dissociation curves (Figs. 3,4,5,6).

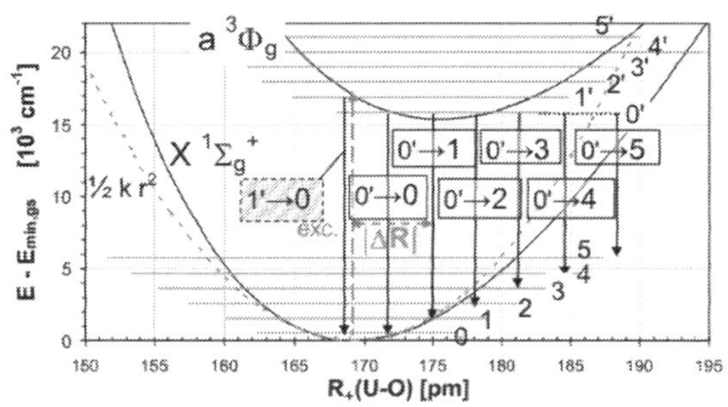

Fig. 1. The symmetric dissociation curves for "bare" uranyl ion UO_2^{2+} calculated by DFT [22-24] (for the ground state, **X** $^1\Sigma_g^+$) and Time-dependent DFT [25-27] (for the excited state[b], **a** $^3\Phi_g$) with B3-LYP functional [28], def-TZVPP basis set [29] and Turbomole V6.5 [30,31] recommended Effective Core Potential [33] for uranium [32].

In Fig. 1, the vibrational states are represented by gray horizontal lines and numbered as suggested in the text above. The R_+(U-O) is the uranyl-oxygen bond distance[c],

[a] A manifold of quasi-degenerated excited states with similar vibrational frequencies and equilibrium geometries is a more realistic description, however, would not change much in later derivation.

[b] The lowest lying triplet state according to the TDDFT/B3LYP/def-TZVPP/ECP-60 method. For *hydrated* uranyl, however, the lowest lying triplet is reported to be $^3\Delta_g$ (which agree with our DFT/B3LYP results) [59]. The true, relativistic symmetry mix triplet state with a singlet component into 1_g state.

[c] Equal for both bonds in the uranyl ion in the symmetric mode.

 DOI: 10.1051/epjconf/201612802002

E electronic energy, $E_{min,gs}$ is (here) an unimportant amount of the equilibrium ground state energy. The thick vertical arrows correspond to the transitions visible in the emission (luminescence) spectrum (Fig. 2), the gray dashed arrow corresponds to excitation transition prior to emission. The absolute value of ground and excited state equilibrium distances, $|\Delta R| = R_{min,es} - R_{min,gs}$ will be discussed later.

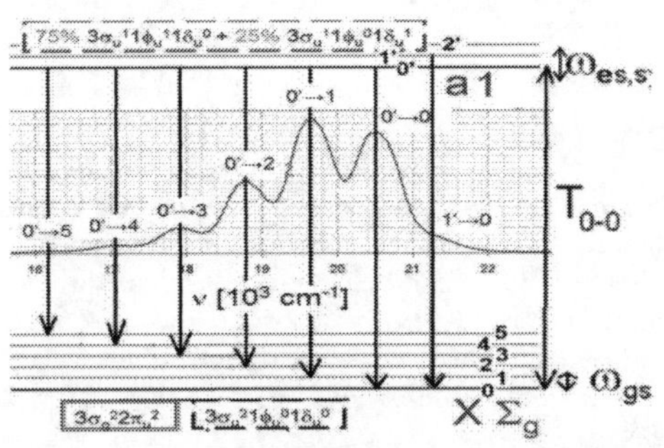

Fig. 2. The pentaaquauranyl ($[UO_2(H_2O)_5]^{2+}$) luminescence spectrum measured by TRLFS in 10^{-4} mol·dm^{-3} U(VI) solution in 4 mol·dm^{-3} $NaClO_4$ and 0.01 mol·dm^{-3} $HClO_4$. Above and below the graph corresponding vibrational levels are shown (see Fig. 1). The vibrational levels of both ground and excited electronic state are spread equidistantly (the eventual anharmonicity of around 5-15 cm^{-1} [34] is within the measurement error) and the spacing corresponds to the ground state Γ_{gs} and excited state vibrational frequencies (870±15 cm^{-1} [35,36,41] and 700±15 cm^{-1} [37] for pentaaquauranyl complex).

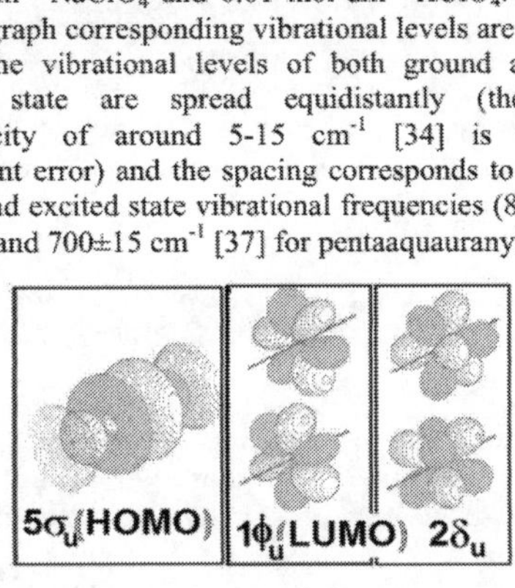

Fig. 3. The isosurfaces of constant positive (light) and negative (dark) wave-function value for the frontier orbitals of the "bare" uranyl ion within the DFT/B3LYP/def-TZVPP/ECP method. The bonding axis O-U-O is drawn as a line for ϕ_u and δ_u orbitals (located almost exclusively on the uranium atom alone), the cylindrical symmetric σ_u orbital is winded around the axis.

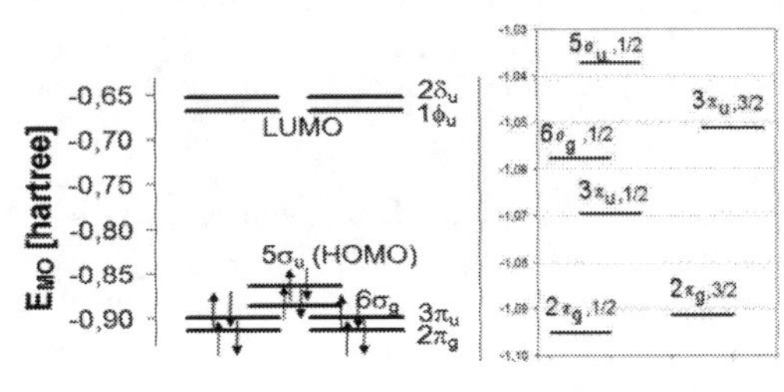

Fig. 4. Diagram of frontier molecular orbitals of the "bare" uranyl ion within the DFT/B3LYP/def-TZVPP/ECP method (left). Exclusion of non-scalar relativistic effects allows double degeneracy of π, ϕ and δ orbitals as well as their orbital angular momenta well-characterisation. Diagram[d] of the highest lying occupied molecular orbitals (or "molecular electronic bispinors") within 4-component all electron DFT derived from Dirac-Coulomb-Gaunt Hamiltonian [21] (right).

Please note the (spin-orbital) splitting of the previously degenerated $3\pi_u$ and $2\pi_g$ energy levels[e]. The ϕ_u and δ_u lowest unoccupied orbitals exhibit similar splitting leading to numerous quasi-degenerated excited states. Under the Dirac-Coulomb(-Gaunt) description [21], the (lowest lying spectral active) excited electronic state is no longer pure triplet, but rather a linear combination of triplet and singlet[f] – the amount of singlet admixture allows for electromagnetic transition to the ground close-shell singlet dominated electronic state. However, we supposed that the energetic properties of the spectral active excited state are similar rather to the triplet than to the singlet state and only excited triplet states are investigated in this contribution. In the Supplementary Information, the singlet states are briefly discussed as well. It is important to stress again, that the true electronic states of molecules in question are linear combinations of components corresponding to different spin multiplicities due to their relativistic nature.

While the symmetric vibrational (stretching) mode of the uranyl group is the most active in spectral resolution, the remaining $3N$-7 vibrational modes[g] will couple as well, but we can think about this coupling as about a mere change from "symmetric vibrational mode (frequency, shift)" to "*effective* symmetric mode (frequency, shift)" [38]. Similarly, the presence of the ligand can be, in the first approximation, seen as a perturbation changing the uranyl electronic excitation energy $T_{00}(0 \rightarrow 0')$ [cm^{-1}], the "effective" symmetric mode vibrational frequencies in the ground state (ω_{gs} [cm^{-1}]) and the excited state (ω_{es} [cm^{-1}]) and the shift of the equilibrium U-O distance due to the excitation (ΔR [pm]).

The T_{00} and ω_{gs} quantities can be obtained from the luminescence peak position fit (Fig. 5). Since the "hot-band" in the luminescence spectra[h] is usually rather wide, the ω_{es} frequency is to be accurately determined from the absorption spectra (Fig. 6) [37].

The ΔR quantity can be determined from the peak surfaces (i.e. the integrals of the spectral intensity corresponding to the individual peaks with the respect to wave-number ν) dependence on vibrational quantum number (i.e. the number of the peak). This is done in Fig. 6 under the preposition that there is only one vibrational mode (the symmetric mode of the uranyl central group) active.

---

[d] The energy axis on right does not correspond to the energy axis on left, since the former is an all-electron calculation result while the latter is ECP variant.

[e] With respect to the Dirac-Coulomb(-Gaunt) hamiltonian, the orbital angular momentum (projected on molecular axis) is no longer good quantum number, so the "σ", "π", "δ" labels are rather for orientation. The projected total angular momentum ("1/2", "3/2" written next to the orbital label) is a good quantum number. The "1/2" are mixture of σ- and π-orbitals, "3/2" of π- and δ-orbitals, etc.

[f] Electronic states are characterized by the total (spin+orbital) angular momentum projected – "J_z = 0" for the ground state and, supposedly "J_z = 1" for the excited state.

[g] N is the number of atoms in the complex molecule (or supramolecular complex of the molecule and solvent molecules)

[h] The transition 1'→ 0 from the vibrationally excited substate of the excited electronic state, which relative intensity grows with the temperature since the initial excited state is thermally equilibrated („thexi") prior to emission.

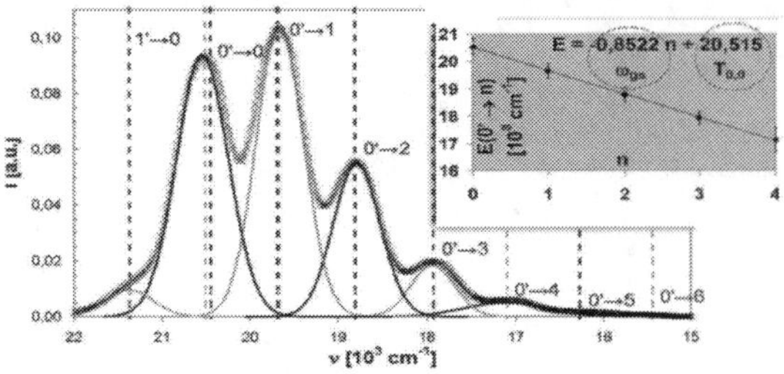

Fig. 5. (the rightmost graph with gray background) The dependence of peak maxima (in cm^{-1}) on the corresponding vibrational quantum number n in the ground state ($0' \rightarrow n$ transition). The main, white background graph shows the pentaaquauranyl luminescence spectrum with wavenumber axis in reversed order[i].

Fitted curve in rigthmost graph in the Fig. 5 corresponds to a single mode linear harmonic oscillators model (the ground state vibrational frequency is a slope of the fitted line and equals to $\omega_{gs} = (850 \pm 20)$ cm^{-1} here), the other parameter of the line fit corresponds to $0' \rightarrow 0$ transition energy T_{00} and was determined as $T_{00} = (20.5 \pm 0.1) \cdot 10^3$ cm^{-1} here.

The spectra (either absorption or emission /luminescence) in the spectral region of interest can be written as a sum of the individual peaks $V_{\Delta n}$,

$$I(\nu) = \sum_{\Delta n = +}^{5} V_{\Delta n}(\nu), \tag{2}$$

where ν is the wavenumber in cm^{-1}, the lower index Δn corresponds to the difference of vibrational quantum numbers, $\Delta n = n - n'$. Each peak can be modelled as Voigt profile ([39,40], gaussian broadened Lorentzian profile)

$$\begin{aligned} V_{\Delta n}(\nu) &= V\left(\nu_{\Delta n}, \sigma_{\Delta n}, \gamma_{\Delta n}, S(\Delta n); \nu\right) \\ &= S(\Delta n) \cdot \frac{\gamma_{\Delta n}}{\sqrt{2}\sigma_{\Delta n}\pi^{3/2}} \int_{-\infty}^{+\infty} \frac{\exp\left(-\left(\nu' - \nu_{\Delta n}\right)^2 / 2\sigma_{\Delta n}^2\right)}{\left(\left(\nu - \nu_{\Delta n} - \nu'\right)^2 + \gamma_{\Delta n}^2\right)} d\nu' \end{aligned} \tag{3}$$

where $\nu_{\Delta n}$ is the peak maximum (plotted in the gray-background graph in Fig. 4 against the $\Delta n \geq 0$, the fit corresponds to $E(0' \rightarrow n) = c\ \nu_{\Delta n} = c\ \omega_{gs}\ \Delta n + T_{00}$), $\sigma_{\Delta n}$ and $\gamma_{\Delta n}$ are width-parameters of Gaussian and Lorentzian part of the peaks respectively and $S(\Delta n)$ determine the peak contribution to spectrum (and is related to the integrated peak intensity via (6)). The Voigt profile can be easily evaluated via the real part of Faddeeva function [42,43] w,

$$V_{\Delta n}(\nu) = \frac{\mathrm{Re}\left[w\left(\dfrac{\nu - \nu_{\Delta n} + i\gamma_{\Delta n}}{\sigma_{\Delta n}\sqrt{2}}\right)\right]}{\sigma_{\Delta n}\sqrt{2}} \tag{4}$$

---

[i] To resemble the characteristic shape of uranyl luminesce spectra when plotted against wavelength) and highlight Voigt profile fit of the individual peaks $V_{\Delta n}(\nu)$ (2).

which is for complex variable z defined via complementary error function erfc. The "i" in formulae (4) and (5) stands for imaginary unit ($i^2 = -1$),

$$\begin{aligned} w(z) &\equiv \exp\left(-z^2\right)\mathrm{erfc}\left(-iz\right) \\ &= e^{-z^2}\left(1 + \frac{2i}{\sqrt{\pi}}\int_0^z e^{t^2} dt\right) \end{aligned}, \tag{5}$$

$$S(\Delta n) = \int_{-\infty}^{+\infty} V_{\Delta n}(\nu)\, d\nu\,. \tag{6}$$

In our simplified model, we adapt Born-Oppenheimer approximation [44] and neglect both nuclear coordinates contribution to transition moment operator and possible dependence of its matrix element between electronic states on nuclear coordinates. As a result, the integrated peak intensities ratio (here with $n'=0$ fixed) equals to ratio of squares of the absolute values of Franck-Condon factors [45-48,50]

$$\frac{S(\Delta n)}{S(0)} = \frac{\left|F(n', n)\right|^2}{\left|F(0', 0)\right|^2}, \tag{7}$$

defined as scalar products of the corresponding n'-th vibrational wave function in the excited state and n-th vibrational wave function in the ground electronic state,

$$F(n', n) = \left\langle \psi_{n'}^{(es)} \middle| \psi_n^{(gs)} \right\rangle, \tag{8}$$

$$\left\langle \psi_{n'}^{(es)} \middle| \psi_n^{(gs)} \right\rangle = \int_{-\infty}^{+\infty} \bar{\psi}_{n'}^{(es)}(R)\, \psi_n^{(gs)}(R)\, dR\,. \tag{9}$$

In the formulae above, n' and n might, in general, stand for arrays of non-negative integers and the integration in (9) might be with respect to all $3N$-6 internal degrees of freedom in the molecule, but within one-mode approximation we work with n', n single non-negative integers and left side of (9) one-dimensional integral. In the harmonic approximation [49] which can be adapted since the uranyl group show, in our experiments, unmeasurable anharmonicity, the corresponding vibrational wave function has a form [48],

$$\psi_n^{(y)}(R) = \frac{\pi^{-1/4}}{\sqrt{2^n n!}} H_n(\xi_y) \exp\left(-\left(\xi_y - \xi_{y,\min}\right)^2 / 2\right) \tag{10}$$

where $y \in \{es, gs\}$ denotes the electronic state, H_n is the n-th Hermitian polynomial [51,52], ζ_y and $\zeta_{y,min}$ are *dimensionless* normal mode coordinate and equilibrium value (i.e. ζ_y and $\zeta_{y,min}$ represent scaled U-O bond distance R and its equilibrium value)

$$\xi_y \equiv R / R_{0,y}, \tag{11}$$

The scaling constants/units are state-specific due to frequency ω_y dependence

DOI: 10.1051/epjconf/201612802002

$$R_{0,y} = \sqrt{/\mu\omega_y}, \qquad (12)$$

where μ is the reduced mass of the respective normal mode (i.e. symmetric vibrational mode of the uranyl group, in case oxygen nuclear masses equals[j], it is multiple of the oxygen nuclear mass, see (40)).

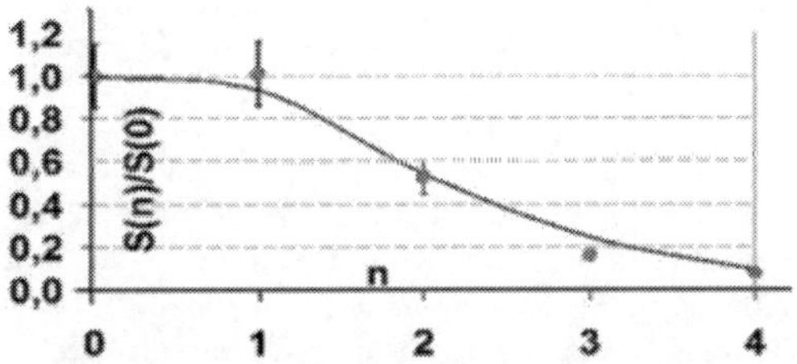

Fig. 6. The ratio of integrated peak intensities S(n)/S(0), defined by (3) corresponding to the 0'→n transitions in the experimental pentaaquauranyl luminesnce spectrum (Fig. 2) – gray filled round circles.

The smooth curve in Fig. 6 above corresponds to the fit to theoretical dependence for a single mode model of two (by ΔR) origin-shifted harmonic oscillators with different frequencies (ω_{gs} and ω_{es})[k]. The fitted [53] value was ΔR = (16±4) pm, which doesn't agree well with the same quantity fitted from the absorption spectrum ΔR = (28±8) pm[l].

Of course, this is for mere classification of the spectral parameters and first-sight interpretation of the spectra, in particular quantum chemical computations, whole (supramolecule) has to be treated as an indivisible system.

2.2 Model of absorption and luminescence signal of a multicomponent aqueous solution

For the absorption solution spectra, we used the multicomponent (additive) variant of the Lambert-Beer law [55] in the form

$$Y_i(\lambda) = \sum_{m=1}^{b} C_{im} Z_m(\lambda) + n_i(\lambda), \qquad (13)$$

where $Y_i(\lambda)$ is the i-th spectrum[m] evaluated at the wavelength λ, C_{im} the molar concentration and Z_m spectrum of the individual chemical species (e.g. UO_2^{2+}, UO_2SO_4, $UO_2(SO_4)_2^{2-}$, ... for the sulphate-system). The total number of distinguishable components is denoted as b. The $n_i(\lambda)$ is the noise function.

The optical length l is incorporated in the $Z_m(\lambda)$ here as an overall multiplicative constant, so precisely,

$$Z_m(\lambda) = l \cdot \varepsilon_m(\lambda), \qquad (14)$$

where $\varepsilon_m(\lambda)$ is the molar absorption coefficient for the m-th species and wavelength λ.

For the TRLFS spectra the same additive law as (13) was adapted, except the $Z_m(\lambda)$ has to be replaced by multidomain function (depending on emission wavelength λ, delay between the excitation pulse and beginning of the luminescence signal integration t, excitation wavelength λ_{exc} and possible other parameters described below). For this case we can write

$$Z_m = Z_m(\lambda, t, \lambda_{exc}, E_{exc}), \qquad (15)$$

where E_{exc} is an excitation pulse energy. We can model the individual component spectra

$$Z_m = S_m(\lambda_{exc}, E_{exc}) \cdot T_m(t) \cdot \overline{Z_m}(\lambda_{exc}, \lambda), \qquad (16)$$

as a products of the energy-dependent saturation part S, temporal part T and pure-spectral part $\overline{Z_m}$. The respective terms can be further modelled [13,61] as

$$S_m(\lambda_{exc}, E) = \mu_m E_{sat,m}\left(1 - \exp\left(-\frac{E}{E_{sat,m}(\lambda_{exc})}\right)\right), \qquad (17)$$

$$T_m(t) = \tau_{m,i}\left(1 - \exp\left(-\Delta t / \tau_{m,i}\right)\right) \cdot \exp\left(-t / \tau_{m,i}\right), \qquad (18)[n]$$

and approximatively

$$\overline{Z_m}(\lambda_{exc}, \lambda) = \varepsilon_m(\lambda_{exc}) \cdot \overline{\overline{Z}}_m(\lambda). \qquad (19)$$

Where E (17) is excitation laser pulse energy, $E_{sat,m}$ is saturation energy (a parameter, only *very slightly* dependent on both m and λ_{exc}). The formula (17) says that increase in excitation laser energy will not yield always the same amount of signal per added mJ. Eventually (almost) all molecules will be excited and the signal cannot reach higher. For very low energies (in praxis for E < 0.7 mJ, the $O(E)$ part of (20) can be neglected with the respect to measurement errors) the energy dependence of S_m and therefore of the whole luminescence signal is linear as can be seen by Taylor expansion of (17),

$$S_m(E) = \mu_m E \underbrace{- \frac{\mu_m}{2E_{sat,m}} + E^2}_{O(E)}. \qquad (20)$$

However, the good signal:noise ratio is achieved for rather higher excitation pulse energies[o] ($E \approx$ 1.5 mJ, or up to $E \approx$

[j] Otherwise, there is no uranyl symmetric stretching mode, just a mode most close to it and its inverse reduced mass will be proportional to sum of inverse masses of both oxygen nuclei.

[k] The theoretical dependence is given by ratio of Franck-Condon factors for the above mentioned simplified model [56] and can be found together with the multi-mode generalization e.g. in [57,58].

[l] Putting the single mode approximation in the serious doubts – mostly for the absorption spectrum, where experimental data and fitted curve disagree to a greater extend.

[m] For an UV-Vis absorption speciation study we can think about i-th solution spectrum. That is the spectrum for a solution in series characterised by total ligand concentration $c_{L,i}$ (unique for the i-th solution) and total uranium concentration c_U, ionic strength I, pH and temperature (those quantities are kept the same for all solutions in the series).

[n] The monoexponential decay is rather approximation and [61] suggest that mostly probably too rough.

[o] In theory, the "saturation area", where signal is almost independent on energy since energy is large enough to excite almost all molecules would be the perfect choice – but it would heat excessively both sample (changing measured properties) and the laser apparatus. On contrary the

4 mJ) and the non-linearity of the dependence of luminescence signal on E has to be taken into account[p].

The temporal dependence T is parametrized by luminescence lift-time $\tau_{m,i}$ of the m-th species in the i-th solution and consist of the time-independent pre-factor resulting from the integration of the signal in CCD camera (integration time Δt) and the supposed mono-exponential decay of the excited state (yet there are suggestions that in some cases, slightly more complicated, e.g. bi-exponential model should be adopted instead). The luminescence life-time is, in general, dependent not only on the luminescence active species (indexed m) but also on the overall composition of the solution, most importantly on the concentration of the quenchers and on the temperature. The simplest, yet relatively accurate model for this dependence is of Stern-Volmer type[q] [1,60],

$$\frac{1}{\tau_{m,i}} = \frac{1}{\tau_{m,i}^{(0)}} + \sum_{q\in Q} k_q [q], \tag{21}$$

where $\tau_{m,i}^{(0)}$ is radiation life-time of the luminescence active excited electronic state (i.e. luminescnce life-time in absence of any quenching species), $[q]$ stands for the molar concentration of quenching specie "q", k_q is the respective rate constant for which we can suppose [36] [r] the Arhenius law dependence on thermodynamic temperature T,

$$k_q = A_q \exp\left(-\frac{E_{A,q}}{RT}\right), \tag{22}$$

where A_q is the frequency factor, $E_{A,q}$ activation energy and R universal gas constant. The term "quenching" is used here loosely – the point is, that some species (H_2O[s], HSO_4^-, $HSeO_3^-$ [9], organic traces [62]) were proven to decrease the $\tau_{m,i}$ value and the respective dependence form can be supposed to be (21) and (22). But the actual mechanism of the non-radiative energy transfer remains unclear[t] [61].

For a constant physical-chemical conditions (pH, ionic strength, temperature, electrolyte type), the $\tau_{m,i}$ increase with decreasing number of water molecules in the first coordination sphere, i.e.

$$\tau_i\left(UO_2^{2+}\right) < \tau_i\left(UO_2XO_4\right) < \tau_i\left(UO_2\left(XO_4\right)_2^{2-}\right) < \\ < \tau_i\left(UO_2\left(XO_4\right)_3^{4-}\right), \tag{23}$$ [u]

where X=S, Se and where the simplified formula for the species in question was putted in round parenthesis instead of being denoted by index m. We suppose that the inequality (23) is due to the similar dependence of $\tau_{m,i}^{(0)}$ on the "intramolecular (coordinated) water concentration", i.e. the $\tau_{m,i}^{(0)}$ and as a result the $\tau_{m,i}$ increase in (23) as the number of water molecules coordinated in the equatorial plane to uranium in molecule decrease with the number of other ligand (XO_4^{2-}) increasing.

The formula (19) says that approximately, the two-wavelength domain luminescence spectrum is a product of an absorption ($\varepsilon_m(\lambda_{exc})$) and emission spectra ($\overline{\overline{Z_m}}(\lambda)$). Since most laser systems in TRLFS measurement are tunable, it is possible to measure both dependence and, in most general case – gain the information from the all three domains (modes, λ_{exc}, λ, t) at the same time [67]. However, the tunning (change in λ_{exc}) is much slower than recording the temporal dependence (change in t) or measurement in the λ-domain (realised by a dispersion element before the CCD camera chip) and therefore, in most of our measurements, λ_{exc} was being kept constant and $\overline{\overline{Z_m}}$ is identified with $\overline{Z_m}$ absorbing the λ_{exc} dependent factor $\varepsilon_m(\lambda_{exc})$ into the multiplicative constant μ_m specific for the particular luminescence active species, but, almost independent on any other parameter or variable.

To distinguish the premultiplicative factor μ_m (17) from other terms, normalization condition on the spectrum in question should be applied. The possible variants include L^2 continuous norm (24), l^2 discrete norm (25) and a maximum norm l^∞ (26),

$$L^2\left[Z_m\right] \equiv \frac{1}{\lambda_{max}-\lambda_{min}}\sqrt{\int_{\lambda_{min}}^{\lambda_{max}} Z_m^2(\lambda)\,d\lambda} = 1, \tag{24}$$

$$l^2\left[Z_m\right] \equiv \frac{1}{N}\sqrt{\sum_{l=1}^{N} Z_m^2(\lambda_l)} = 1, \tag{25}$$

$$l^\infty\left[Z_m\right] \equiv \max_{\lambda\in\langle\lambda_{min},\lambda_{max}\rangle} Z_m(\lambda) = 1. \tag{26}$$

where λ_{max} and λ_{min} are boundaries of the spectral region of the interest, N is the number of spectral points, $\lambda_1 = \lambda_{min}$, $\lambda_N = \lambda_{max}$.

2.3 UO_2^{2+} - XO_p^{2-} - H_2O speciation and Thermodynamic properties of uranyl complexes

The uranyl complex molecules in question consist of (none, one, two or three) XO_p^{2-} ligands coordinated to the uranium atom in the linear O≡U≡O group[v] in the equatorial plane (plane perpendicular to the uranyl group,

"linear response area" ($E << 0.7$ mJ) would be best for deconvolution of the energy dependence, but the signal: noise ratio would be low.

[p] It is ideal to keep the settings of the laser system such that E vary as less as possible, but rarely it is possible to keep relative square root variations under 5%. And therefore we cannot simply incorporate energy dependence into some pre-multiplicative constant.

[q] Could apply for both intermolecular quenching (collisional, diffuse) and intramolecular quenching (Förster, Dexter), difference will be in the k_q interpretation (see Supplementary Info).

[r] Please see Fig. 21 and 22 on 13th page of [36].

[s] For example, in the system UO_2^{2+} - $NaClO_4$ – $HClO_4$ – H_2O, for constant *pH*, the higher is ionic strength, the longer is observed $[UO_2(H_2O)_5]^{2+}$ luminescence life-time. The luminescence life-time ionic strength dependence could be explained by equation (21) with single quenching species, q = H_2O, since higher ionic strength imply lower molar concentration of water in the water solution. First author's preliminary measurements [36] (Fig. 20) are in agreement with that fit.

[t] To cite [61] exactly "*The nature of aqueous uranyl deactivation is still under discussion, and both photochemical and physical mechanisms of excided state quenching are proposed.*"

[u] Well documented in first table in [63]

[v] For uranium-oxygen triple bond character – [66].

DOI: 10.1051/epjconf/201612802002

[2]). In case of no ligands, the hydration number of uranyl group was determined as 5 [2,64,65], i.e. pentaaqua uranyl complex $[UO_2(H_2O)_5]^{2+}$, but with ligands, the coordination number of the uranyl group can be either 5 or 6 [2]), also sulfate, selenate and carbonate ligands can coordinate by either one oxygen donor atom (κ^1-, monodentate coordination) or by two oxygen donor atoms (κ^2-, bidentate coordination). As schematically shown in Fig. 7.

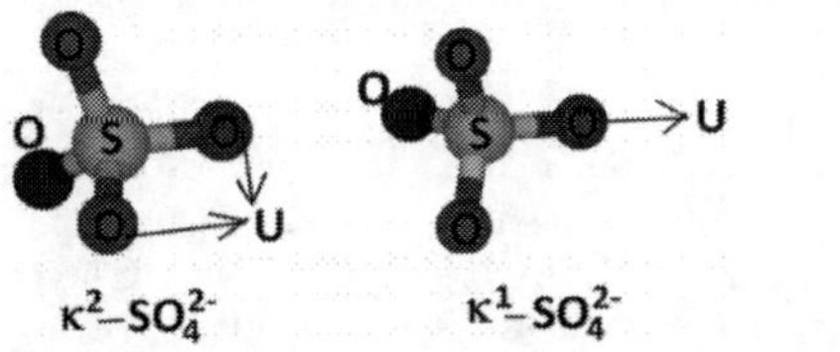

Fig. 7. Schematic view of bidentate (κ^2-) and monodentate (κ^1-) coordination of sulfate ligand to uranium atom. The ligand has two (κ^2-) or one (κ^1-) donor oxygen atom available.

For 1:1 sulfate complex this leads to four possibilities (the simplified description in the round parenthesis describe ligand bonding and uranyl coordination number) -$[UO_2(\kappa^2\text{-}SO_4)(H_2O)_3]$ (κ^2co5), $[UO_2(\kappa^2\text{-}SO_4)(H_2O)_4]$ (κ^2co6), $[UO_2(\kappa^1\text{-}SO_4)(H_2O)_4]$ (κ^2co5) and $[UO_2(\kappa^1\text{-}SO_4)(H_2O)_5]$ (κ^2co6), while for 1:2 sulfate complex there are six possible molecular geometries in total (κ^{22}, κ^{12} and κ^{11} ligand bonding with either co5 or co6 total coordination number) and for 1:3 sulfate complex eight (!). In reality we can suppose that there will be only the most symmetric variant for the uranyl tris(sulfate) – $[UO_2(\kappa^2\text{-}SO_4)_3]^{4-}$ (κ^{222}co6) and that among the six isomers/conformers of uranyl bis(sulfate) and four of uranyl monosulfate only some will be energetically favoured and therefore measurably present in the solution. In particular, the combination of κ^1-bonding and higher coordination number co6 would be disfavoured. But both experimental data [68] and theoretical quantum chemical calculations [68,69] suggests that more than one, possibly more than two isomers are present together for both mono- and bis- sulfates and the same possibly applies for selenates.

In the rest of this section the coordinated water molecules will be omitted for simplicity, yet the change of the water molecules number in the reaction stechiometry still kept in mind. Similarly, the κ^1-/κ^2-coordination mode isomery is omitted. In fact, to all isomers discussed above might be assigned individual stability constants (29)-(32), SIT-interaction parameters $\varepsilon(i,j)$ (33) (where i stands for the luminescence active species and j for main electrolyte ionic species – Na^+, ClO_4^-, SO_4^{2-}, HSO_4^-, H_3O^+) and spectra. By precise chemometric analysis of multiple solution series (differing by ionic strength and *pH*, possible also c_U), possibly for several temperatures it might be possible to distinguish two or more isomers of complex with the given stechiometry. For example to distinguish $[UO_2(\kappa^2\text{-}SO_4)(H_2O)_3]$ (f_m = 2, see (27), (31) and (33) below) from $[UO_2(\kappa^1\text{-}SO_4)(H_2O)_4]$ (f_m = 1). Yet only EXAFS study was able to do this reliably [68]. The UV-vis or TRLFS study would need an uninterrupted several weeks lasting continual access to a very stable measurement system – which we didn't achieve yet.

The aqueous uranyl - sulfate (UO_2^{2+} - SO_4^{2-} - H_2O) system is characterised by chemical equilibria

$$UO_2^{2+} + m\,SO_4^{2-} \quad UO_2(SO_4)_m^{2-2m} + f_m\,H_2O, \quad (27)$$

$$HSO_4^- + H_2O \quad SO_4^{2-} + H_3O, \quad (28)$$

characterized by concentration-defined ("concentrational") equilibrium "constants",

$$\beta_m = \frac{\left[UO_2(SO_4)_m^{2-2m}\right]}{\left[UO_2^{2+}\right]\left[SO_4^{2-}\right]^m}, \quad (29)$$

$$K_2 = \frac{\left[SO_4^{2-}\right]}{\left[HSO_4^-\right]}\cdot 10^{-pH}, \quad (30)$$

where $[X]$ stands for the molal ($mol\cdot kg_w^{-1}$, i.e. moles in 1 kg of the water) concentration of the specie X, $m \in \{1, 2, 3\}$ and f_m is a parameter discussed in later text. The true, thermodynamic constants β_m^o, K_2^o are defined through chemical activities of all species present in the chemical reaction. Denoting $a(X)$ the chemical activity of specie X,

$$\beta_m^o \equiv \frac{a\left(UO_2(SO_4)_m^{2-2m}\right)a(H_2O)^{f_m}}{a\left(UO_2^{2+}\right)a\left(SO_4^{2-}\right)^m}, \quad (31)$$

$$K_2^o = \frac{a\left(SO_4^{2-}\right)}{a\left(HSO_4^-\right)}\cdot 10^{-pH}. \quad (32)$$

The relation between the two kind of constants, i.e. the dependence of the concentrational "constants" (29)-(30) on the truly constant thermodynamic parameters (31)-(32) and concentration of all species present in the solution can be modelled by Specific Ion Interaction Theory (SIT) [70-72] via equations ("SIT equations"),

$$\log\beta_m = \log\beta_m^o + z_m^2 D(I)\Delta \sum_k (\varepsilon-)_{mk}[k]\Delta f_m \log + a_w$$

$$\log K_2 = \log K_2^o + 3D(I) - \sum_k (\Delta\varepsilon)_{0,k}[k] + f_0 \log a_w, \quad (33)$$

where aw is water activity[w] $[k]$ is the molality of specie "k", $\Delta\varepsilon_{mk}$ is a linear combination of SIT interaction parameters $\varepsilon(i,j)$, Δz_m^2 = -2, -2, 0 for m = 1, 2, 3, $D(I)$ is a Debye-Hückel term ([73,74], a function of ionic strength[x] of a form (34))

$$D(I) = \frac{A\sqrt{I}}{1+B\sqrt{I}}, \quad (34)$$

where constants A = 0.509 $kg^{1/2}\cdot mol^{-1/2}$ and B = 1.5 $kg^{1/2}\cdot mol^{-1/2}$ for 25°C for their (slight) temperature dependence see [74]. The SIT equations are in some cases

[w] In the simplest case function of chosen electrolyte, I and temperature T.

[x] Ionic stength I is defined as $I = \frac{1}{2}\Sigma_k[k]q_k^2$, where $[k]$ is a molal concentration and q_k charge-number of k-th specie in the solution.

simplified into an approximative form [74], where β_m depends on concentrations only through ionic strength, i.e.

$$\log \beta_m = \log \beta_m^o + z_m^2 D(I)\Delta \varepsilon_m I - f_m \Delta \log a_w . \quad (35)$$

Together with the bilantion equations,

$$c_U = \sum_{m=0}^{3}\left[UO_2(SO_4)_m^{2-2m}\right], \quad (36)$$

$$c_S = \left[SO_4^{2-}\right]+\left[HSO_4^-\right]+\sum_{m=1}^{3} m\left[UO_2(SO_4)_m^{2-2m}\right], \quad (37)$$

where left sides are total uranium and sulphate molalities, the SIT equations (33) fully determine the speciation model, i.e. the dependence of the model concentration of m-th spectral active specie on the concentration parameters of the i-th solution,

$$C_{im} = C_{im}\left(\beta, c, c_{S,i}\right), \quad (38)$$

Where $c_{S,i}$ is the total sulfate molal concentration in the i-th solution, c stands for all other (i-th) solution parameters and β stands for the stability constants (31), (32) and $(\Delta\varepsilon)_{mk}$ SIT parameters (33), or for stability constants (29)-(30) in the case of the simplified model (35). The parameters β are then determined through least-square (in general maximum likelihood) based chemometrical method from the experimental spectral data together with the spectral parameters discussed in previous section 2.2.

The speciation is temperature dependent mostly through temperature dependence of the stability constants. We limited ourselves to the case of temperature independent reaction enthalpies ΔH_m^o and entropies ΔS_m^o (for the studied temperature interval 10°C - 45°C) a therefore Van't Hoff isotherm [20] can be used for the description,

$$\ln \beta_m^o = -\frac{\Delta H_m^o}{RT} + \frac{\Delta S_m^o}{R}. \quad (39)$$

2.4 Quantum chemical description of uranyl complexes in aqueous solutions

We ought to study eigenvalues and eigenstates of molecular hamiltonians corresponding to the hydrated species in question – this is done directly through the true *ab initio* (wave-function based) methods such as Hartree-Focks method (HF, SCF), post-Hartree Fock methods (Moller-Pleset perturbational MP2 method, limited approximate configuration interaction CIS(D) [81] and CC2 as the simplest method from the coupled-clusters family) [24]. In contrast to the wave-function based methods, there is also, possibility to access the energy level values through equations for electronic density – by Density Functional Methods (DFT) [22-24]. Though the DFT could theoretically yield results fully including electron correlation with a very low computational costs (similar to HF), the problem is that the yet known density functional are only approximate. In particular, the in chemistry most widely used, hybrid functional B3-LYP (Becke, three-parameter, Lee-Yang-Parr, [28,82-85]) had his three empirical parameters fitted to formation enthalpies of molecules with light atoms and its usability for uranium containing systems is yet in question (however, referring to previous positive tests of B3-LYP for uranyl systems, [75-80,34], we can suppose the methodics is at least accurate for qualitative analysis).

To be able to exploit the resolution-of-identity technique (significantly accelerating computations) in Turbomole V6.5 [30,31], we used the GGA (Generalized Gradient Approximation), non-hybdrid, P-B[y] functional as well.

The calculations mentioned above were done by Turbomole V6.5 program [30,31] with only part of scalar relativistic effects included via the Effective Core Potential (ECP, [32,33]). In this approach, the inner 60 electrons[z] on the uranium atom are included by an effective potential in which the other electrons motions are described by non-relativistic Schrodinger equation (or in particular Restricted Hartree-Fock equations + post-HF treatment as written above or the Restricted Kohn-Sham equations of DFT). The effective core potentials also allow for approximative inclusion of finite-size nuclear effects and simplify electronic calculations in general (reducing the number of electrons treated explicitly).

Only few calculations were done via all-electron four-component DFT/B3LYP based on Dirac-Coulomb-Gaunt (DCG) hamiltonian [21] via the DIRAC 13 software [88] (e.g. the molecular orbital diagram detail showing splitting in the right part of Fig. 4). The splitting of molecular orbitals is a vector relativistic effect and cannot by included by scalar ECP method described in previous paragraph. The Dirac-Coulomb (DC) or DCG hamiltonians are essential to description of the true symmetry properties of ground and excited state wave-functions (which are, e.g. for free UO_2^{2+} now eigen-functions of J_z – the total (spin+orbital) angular momentum projected on molecular axis and not eigen-function of either the orbital L_z or spin s_z part alone) and calculation of electronic transition moments and therefore estimation of the μ_m quantity from (17) as well as the norm of the absorption spectra (while for its shape, Franck-Condon factors based on scalar ECP results are sufficient).

For the hydration modelling we tried to use the simple variant of polarized continuum model (COSMO) incorporated in Turbomole V6.5 [89], but abandoned this approach after failing to achieve meaningful values of thermodynamic properties - β_m^o = - RT ln ΔG_m^o, ΔH_m^o and ΔS_m^o. Therefore, more precise method of explicit hydration was adapted. From first estimates on uranyl symmetric stretching mode vibrational frequency in both ground and excited electronic state and T_{00} excitation energy which compares well with experiment, we believe

[y] Becke-Pedrew (B-P) functional consists of Exchange part :S[82,83]+B88[86], Correlation part:VWN(V)[84]+P86[87].

[z] most influenced by the relativistic effects due to the higly charged uranium nuclei

that one complete sphere of water molecules[aa] fully surrounding the complex in question (forming together supramolecular complex) might be enough at least for spectroscopic properties estimation. Requirement on minimizing the computational artefacts connected to the surface tension on the artificial supramolecular complex boundaries lead to need to use spherical supramolecules with up to 65 solvating water molecules inside. Within RI-DFT [90,91] or, perhaps, (MA)RI-MP2 methods [92] in Turbomole this is still feasible – yet we are still waiting for the first results of comptations of this kind.

While we used up to def-TZVPP atomic basis set [29] for complex molecule atoms, to the hydration sphere water molecule we assigned the def-SVP [30,100] atomic basis set [90] only, to keep the computation time realistic.

For accurate description of the explicit solvation, the dispersion interactions and correlation energy are crucial. Yet, the empirical dispersion corrected DFT-D3 was used [93], but later we should try comparison with e.g. MP2, MP2(F12) or MP4, MP4(F12) methods for supramolecular complexes of smaller to higher size.

2.4.1. Quantum chemical estimation of spectroscopical properties

All 3*N*-6 vibrational modes for both singlets ground state and the first excited triplet state can be calculated via GF method [94] from the ground and excited state Hessian matrix (i.e. matrix of all second derivatives of electronic energy with the respect to nuclear coordinates). While we were interested here mere in the one or two modes most closely resembling the symmetric stretching of the uranyl central group (with frequency ω_{gs} and ω_{es} – for the ground and excited state respectively), the other modes may play significant role in the luminescence and absorption spectra shape prediction as well (through the Franck-Condon factors [45-50]).

For the free uranyl ion, UO_2^{2+}, for example, the *symmetric mode stretching frequency* ω_y in electronic state $y \in \{gs, es\}$ equals[bb],

$$\omega_y = \sqrt{\frac{1}{\mu}\frac{\partial^2 E_y}{\partial R^2}}, \quad (40)$$

Where E_y is electronic energy and in case R is average of both U-O bond lengths, the reduced mass μ equals double[cc] of the oxygen nuclear mass (under assumption oxygen nuclear masses equals). Let as define the *adiabatic excitation energy* T_a as a difference of minima $E_{es,min} = \min_R E_{es}(R) = E_{es}(R_{es,min})$ on the excited state potential energy surface (PES, $E_{es}(R)$) and the ground state PES minima $E_{gs,min} = \min_R E_{gs}(R) = E_{gs}(R_{gs,min})$,

$$T_a = E_{es,min} - E_{gs,min}, \quad (41)$$

The 0' → 0 transition energy T_{00} is connected to the adiabatic excitation energy as

$$\begin{aligned} T_{00} &= \left(E_{es,\min} + ZPE_{es}\right)\ \left(-E_{gs,\min} + ZPE_{gs}\right) \\ &= T_a + ZPE\,\Delta \end{aligned}, \quad (42)$$

where $ZPE_y = \frac{1}{2}\sum_{n=1}^{3N-6}\omega_{y,n}$ is the zero-point energy in the electronic state y.

Vertical excitation energy is defined as a difference between excited state energy and ground state energy both at the same geometry $R_{gs,min}$ corresponding to the ground state equilibrium,

$$T_{v,ex} = E_{es}\left(R_{gs,min}\right) - E_{gs}\left(R_{gs,min}\right), \quad (43)$$

roughly, this energy corresponds to the highest peak maximum position in the absorption spectrum.

Vertical deexcitation energy is defined analogically as an electronic states energy difference at the excited state equilibrium geometry $R_{es,min}$,

$$T_{v,d} = E_{es}\left(R_{es,min}\right) - E_{gs}\left(R_{es,min}\right), \quad (44)$$

roughly corresponding to the peak maximum position of the highest peak in the luminescence/emission spectrum.

The shape of the luminescence and absorption spectra are determined by Franck-Condon factors, which are function of vibrational frequencies in the both electronic states and the difference of equilibrium geometries in the excited and the ground state,

$$\Delta R = R_{es,min} - R_{gs,min}, \quad (45)$$

and mutual orientation of excited and ground state 3N-6-tuples or normal mode vectors, for bare uranyl group, this reduce to the difference of U-O equilibrium bond lengths in excited and ground state (formula (45) without vector arrows). From spectral data, we can fit only the absolute value, $|\Delta R|$, but all reasonable theoretical calculations suggest, that $\Delta R > 0$, i.e., the uranyl group bond is elonged due to the electronic excitation (in accordance with the fact, that excitation populate the non-bonding LUMO[dd]

[aa] 30 to 40 water molecules depending on the particular complex species.

[bb] Under the assumption that the two equilibrium U-O bond lengths are identical for given electronic state (i.e. the linear uranyl ion has a center of symmetry in uranium nuclei for both electronic states in question). The assumption is well fulfilled for both bare and coordinated uranyl.

[cc] In case we work in normalized coordinates,

$$R = \frac{1}{\sqrt{2}}\left(\left(r_{O(1)} - r_U\right) - \left(r_{O(2)} - r_U\right)\right)\ \frac{1}{\sqrt{2}} = \left(r_{O(1)}\ \ r_{O(2)}\right) -$$

i.e. $\sqrt{2}$ - multiplied R considered in the text, the reduced mass has to be half of that considered in the text, i.e. the oxygen nuclear mass.

[dd] lowest unoccupied molecular orbital, en.wikipedia.org/wiki/HOMO/LUMO

from bonding HOMO[ee] and therefore decrease effective bond order).

The spectral norm (i.e. the overall luminescence or absorbance of unit concentration), not just the shape of the spectra, represented in case of luminescence by μ_m factor is proportional to the electronic transition probability,

$$\Gamma = \frac{2\pi\rho}{\hbar}\left|\left\langle\psi_{es}(r)\middle|\hat{T}\middle|\psi_{gs}(r)\right\rangle\right|^2, \qquad (46)$$

where ρ is energy density of emitted or absorbed el.mag. field, the transition operator $\hat{T}$ commute with the spin operators and therefore, for pure triplet-singlet radiative transition has exactly zero probability $\Gamma = 0$, i.e. is forbidden. We cannot compute Γ or μ_m within the scalar-relativistic description. When spin-orbit effects are taken into account, the excited state will have also a small singlet admixture (and the ground state similarly even smaller triplet admixture) allowing for non-zero Γ (equal to the matrix element between the singlet components of both states plus triplet-triplet part). We plan to exploit the DIRAC program [88] capabilities to calculate Γ in further studies.

The *widths* or *width-parameters* of individual peaks in spectra might be determined (but were not yet) by two different methodics –

1. By an exhaustive search of all possible supramolecular configurations and the corresponding ground state PES minima (and to them closest lying excited PES minima) and subsequent Boltzmann-factor weighted averaging of the obtained results (spectral peaks for individual configuration might be lorentzian with natural width and possible scaled Doppler Gaussian broadening)
2. By molecular dynamics (MD) study (calculating vertical excitation energy along MD trajectory in the ground state for absorption spectrum and vertical deexcitation energy along MD trajectory in the excited state for luminescence spectrum).

Inclusion of non-adiabatic coupling in "2." (MD study in excited state) may allow for non-quenched luminescence life-time estimation as well. Incorporation of "full molecular dynamics" treating electrons and (light) nuclei on equal footing would allow for modelling of the quenching process and determination of k_q (or in particular A_q and $E_{A,q}$) values as well. The luminescence life-time modelling would be an ambitious plan even for far future.

2.4.2. Quantum chemical estimation of thermodynamic properties

The calculation is based on partition sum Q based statistic physics prediction for Helmholtz energy F, entropy S and internal energy U of an ideal-gas-like system of N identical supramolecular complexes [95]

$$F = -k_B T \ln Q, \qquad (47)$$

$$U = k_B T^2 \left(\frac{\partial \ln Q}{\partial T}\right)_V, \qquad (48)$$

$$S = (U - F)/T, \qquad (49)$$

where k_B is Boltzman constant and T thermodynamic temperature. From the definition of standard molar reaction properties and neglecting the change of pV (pressure x volume product) in course of chemical reaction (then $\Delta G° = \Delta F°$, $\Delta H° = \Delta U°$) we can write

$$\Delta G^o = E_0\Delta \quad ZPE + \Delta G_{vib} + RT\left(\ln N_A \Delta 1\right) + n \qquad (50)$$

$$\Delta H^o = E_0\Delta + H_{vib}, \qquad (51)$$

$$\Delta S^o = \left(\Delta H^o - \Delta G^o\right)/T, \qquad (52)$$

when in (50)-(51) using usual simplifications – electronically excited states are not populated under investigated temperatures, Born-Oppenheimer approximation, rigid rotor and harmonic oscillator description for vibrational motions. Contribution of rotational and translational degrees of freedom of supramolecular complexes to partition sums[ff] are neglected as well. Quantities on right sides of (50) and (51) are defined below for general chemical reaction with stochiometric coefficients ν_i (negative for reactants, index i correspond to chemical species in the reaction)

$$\Delta X = \sum_i \nu_i X_i, \qquad (53)$$

with X standing for E_0 (ground state electronic energy in equilibrium geometry (of i-th species, this index is suppressed in the later text)), ZPE (zero-point vibrational energy in the ground electronic state), G_{vib} (vibrational contribution to $\Delta G°$, (54) below), H_{vib} (vibrational contribution to $\Delta H°$, (55) below),

$$G_{vib} = RT\sum_n \ln\left(1 - \exp\left(-x_n\right)\right), \qquad (54)$$

$$H_{vib} = RT\sum_n \frac{x_n \exp\left(-x_n\right)}{1 - \exp\left(-x_n\right)}, \qquad (55), [95]$$

with

$$x_n = \frac{\omega_n}{k_B T}, \qquad (56)$$

where n index the vibrational modes of i-th species, ω_n is the corresponding vibrational frequency.

When we take $X = 1$, the ΔX corresponds to the change of the number of moles in the reaction, Δn. The last term in (50) is a result of $Q = q^N/N!$ for partition sum Q of N indistinguishable systems, each with partition sum q, for the case $N = N_A$ (Avogadro constant).

The general idea was to consider either the complexation reaction

[ee] highest occupied molecular orbital

[ff] since the corresponding motions are partly blocked in liquid matter and the contributions would vanish in limit of complex specie surrounding solvent molecules number → ∞

DOI: 10.1051/epjconf/201612802002

$$\left[UO_2\left(H_2O\right)_5\right]^{2+}\cdot r_1H_2O \quad + \quad n\,XO_u^{2-}\; r_2H_2O$$

$$\left[UO_2\left(\kappa^{\alpha}-XO_u^{2-}\right)_n\left(H_2O\right)_{K-A}\right]^{2(1-n)}\cdot r_kH_2O,\quad(57)$$

$$+ \quad \Delta r\; H_2O$$

or

$$\left[UO_2\left(H_2O\right)_5\right]^{2+}\cdot s_1H_2O \quad + \quad n\,HXO_u^{-}\; r_2H_2O$$

$$\left[UO_2\left(\kappa^{\alpha}-XO_u^{2-}\right)_n\left(H_2O\right)_{K-A}\right]^{2(1-n)}\cdot s_kH_2O\;.\quad(58)$$

$$+ \quad \Delta s\; H_2O \qquad n\,H_3O^{+}\;\; p\,H_2O$$

The latter consider as a reactant protonated ligand HXO_u^- (X=S, Se ($u = 4$), X = C ($u = 3$)) to avoid multiply charged anion computation (which demands peculiar atomic basis with enough diffuse functions due to loosely bounded HOMO electrons) at least for the case $n = 1$. The number of explicitly accounted hydration water molecules in supramolecules in (57) and (58) have to fulfil

$$r_k \;=\; r_1 + n r_2 + 5 \;\; K - A + \;\; r, \qquad (59)$$

$$s_k \;=\; s_1 + n\left(s_2 - \;\; p\right) + 5 \;\; K - \; A \; + \; s, \qquad (60)$$

where K is the coordination number of uranyl in complex, $A = n_1 + 2\,n_2$ is number of donor atoms of all ligands[gg], $n = (n_1,n_2)$ are numbers of monodentate- and bidentate- (Fig. 7) coordinated ligands[u] ($n = n_1 + n_2$). The Δr H_2O and Δs H_2O stands for one supramolecule of Δr water molecules, i.e. the stechiometric coefficients in (57) and (58) are (-1, -n, 1, 1) and (-1, -n, 1, 1, 1) respectively. While in the limit of all r_j or $s_j \rightarrow \infty$, their particular ratio should not matter, for finite, computation-feasibly small values of r_j, s_j, their ratio will be important. We suggest to compute „correction to artificial surface tension change"(in the modelled water droplets around the species of interest) by subtracting thermodynamic property of chemical reaction where only surrounding water molecules are considered (without the solute). To minimize this correction, the r_j or s_j should be for $n = 1$ chosen as equal and for $n > 1$ we should divide the Δr into n equal portions to compute the reaction with the equal number of supramolecular complexes of equal size on each reaction side.

The stability constants are related to the molar standard reaction gibbs functions as

$$\ln\beta_m^o \;=\; -\Delta G_m^o\,/\,RT\;. \qquad (61)$$

The ΔG° for the reaction (58) is related to ΔG° for (57) as

$$\Delta G_{57}^o \;=\; G_{58}^o\,\Delta \qquad + G_2^o, \qquad (62)$$

where $\Delta G_2^o = \exp\!\left(\;K_2\,/\,RT\right)$, K_2 being second dissociation constant of H_2XO_u acid (adopted from literature/experimental data).

3 Experimentals

3.1 UV-VIS spectrophotometry

For spectrophotometric measurements 0.25 M UO_2^{2+} in 0.42 M $HClO_4$ solution (prepared by dissolving solid UO_3 (p.a., depleted, n.p. Brno (Chemapol, Praha-Czechoslovakia), ≥ 98%) in conc. $HClO_4$ and subsequent demi-water dilution) have been used. Na_2SO_4 and Na_2SeO_4 as a ligand sources, $NaClO_4.H_2O$ (all three – p.a., Sigma-Aldrich, ≥ 98%) to adjust ionic strength, 1 M $HClO_4$ (by dilution of 60% p.a., Fluka) to adjust pH and demineralized water (Milipore, Direct-Q UV3, 18.2 MΩ.cm (25°C)).

For uranyl-selenate system spectrophotometric measurement, higher purity $HClO_4$ has been used (Sigma-Aldrich, 70% conc., ≥ 99.999% purity, SHBB3741).

The measurements were done by Varian (Cary 100 Conc) spectrophotometer (Fig. 8) in wavelength range 381-481 nm (step 1 nm) for temperature (23.4 ± 0,5) °C (uranyl-sulfate system) and temperatures 17.5°C, 20.0°C, 22,5°C, ..., 35°C; 40°C; 45°C (± 0,5°C) (uranyl-selenate sys.). The cuvette holder (Varian) was thermostated by the Peltier effect.

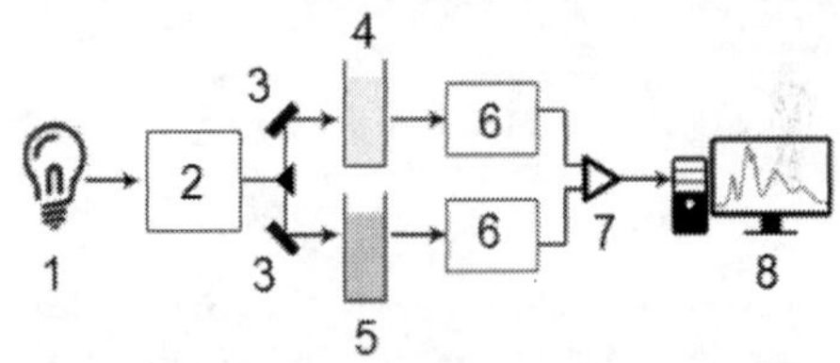

Fig. 8. The UV-vis absorption (spectrophotometric) measurement system. 1 – electromagnetic radiation source, 2 – monochromator, 3 – beam splitter, 4 – reference cuvette, 5 – cuvette with the measured solution, 6 – detector, 7 – amplifier, 8 – signal processing unit.

3.2 TRLFS spectroscopy

For TRLFS measurements, the same chemicals as for spectrophotometry has been used, but the sodium salts were of higher purity (≥ 99.99%, Sigma-Aldrich).

The measuring system is described in Figure 9, the laser (2) generates 1064 nm pulses with 4 ns duration and 10 Hz repetition frequency. Excitation pulse output wavelength 266 nm was generated by double frequency dubling on crystals with $\chi(2)$ non-linearity (SHG 532 nm, SHG 266 nm). The 600 mm^{-1} optical grid (spectral range 163 nm) (7) have been centred at 510 nm. The ICCD camera integration time was set $\Delta t = 10$ μs.

The uranyl-sulfate system was measured for ionic strength $I_M = 0.3$ mol·dm^{-3} for $T = 18$°C and $pH = 2$, further details can be seen in (bakalarka).

For the uranyl-selenate system measurements for ionic strength $I = 1.1$ mol·Kg_w^{-1} were done in Prague at Department of Nuclear Chemistry, energy of the excitation

[gg] except for water

pulses was[hh] 1.6 ± 0.2 mJ, excitation wave-length λ_{exc} = 415 nm. Measurements for ionic strength I = 0.05 mol·Kg_w^{-1} were done in INP Orsay in collaboration with Dr. Vladimir Sladkov, excitation wave-length was 430 nm and average pulse energy E = (4.0 ± 0.5) mJ (energy of individual pulses was not measured in this case).

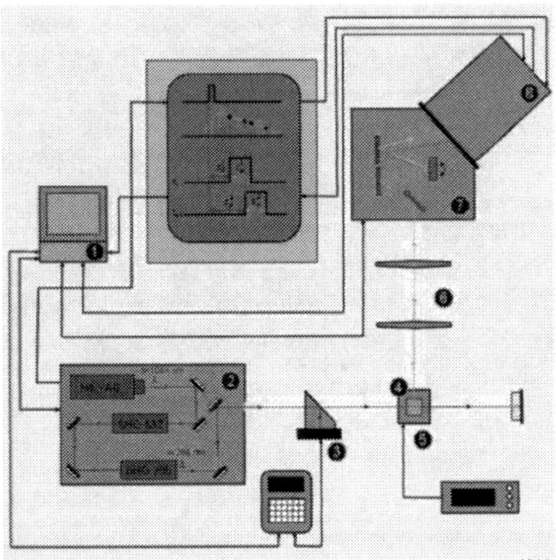

Fig. 9. The TRLFS measurement system stationed at Prague, Nuclear Chemistry Department. 1 – PC control unit, 2 – tunable Nd:YAG laser pumped laser system VIBRANT™ 355 II, 3 – beam splitter diverting defined part of the beam into digital time-resolved joulemeter FieldMax II, which records individual pulses energies and the 1 synchronize the information with the CCD camera measurement results, 4 – quartz cuvette (SUPRASIL) with the solution in thermostated holder (5). 6 – collimating lenses, 7 – Spectrograph MS257TM, dispersion element (optical grid), 8 – ICCD camera ANDOR iStar with time resolution.

4 Experimental results

4.1 UV-VIS spectrophotometry, uranyl – selenate

For the uranyl-sulfate system, please see [7] and [96,36], where values of stability constants have been corrected.

Uranyl-selenate system has been measured for relatively high ionic strength I = (3.8 ± 0.3) mol·Kg_w^{-1} in order to allow for SeO_4^{2-} ligand analytical/total concentrations up to 1.36 mol·Kg_w^{-1} opening the possibility of tris(selenate)uranyl, $UO_2(SeO_4)_3^{4-}$, to be first time studied. *pH* of the 20 measured solutions varied from 1.6 to 2.5 (but was measured with error less than 0.05 for each solution). The wave-length range was 350-500 nm (step 0.5 nm).

The normalized individual component absorption spectra (Fig. 10) show typical bathochromic shift with the respect to the increasing number of SeO_4^{2-} coordinated to the central uranyl group. The speciation diagram (Fig. 11) presents molar concentrations of individual spectral active species as a function of ligand (SeO_4^{2-}) total molality.

[hh] The variability of 0.2 mJ is a result of laser power variation, but energy of each pulse has been measured with roughly ten times higher precision.

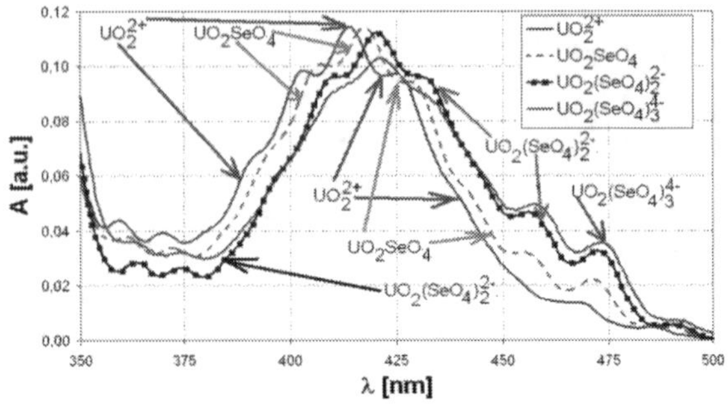

Fig. 10. The normalized[ii] individual component absorption spectra for 25°C derived by PCA-based data analysis of complex solution spectra for uranyl-selenate system.

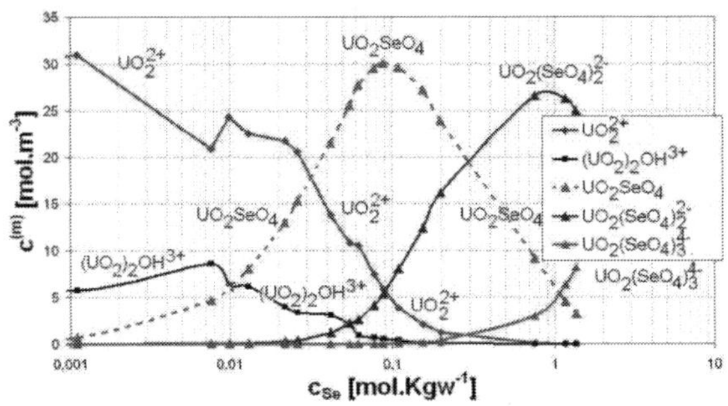

Fig. 11. The model speciation diagram for the fitted (Tab. 1) stability constants for 25°C.

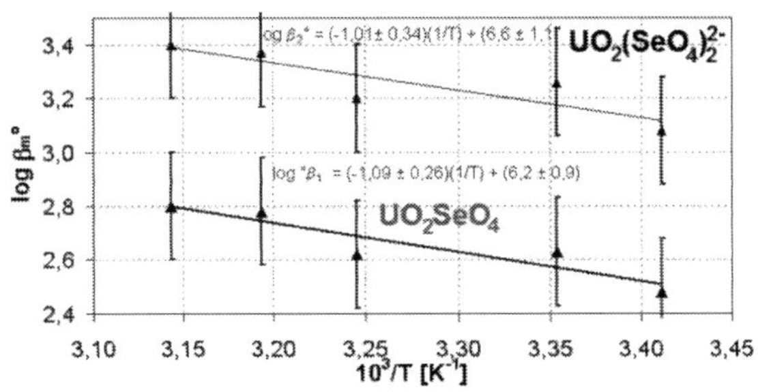

Fig. 12. The Van't Hoff equation fitted thermodynamic stability constants temperature dependence.

By fitting the thermodynamic stability constants (Tab. 1) to Van't Hoff equation (Fig. 12), complexation standard enthalpies and entropies were derived (Tab. 2).

Shapes of absorption spectra for $UO_2(SeO_4)_m^{2-2m}$ and $UO_2(SO_4)_m^{2-2m}$ are identical within the measurement error (for given value of m), e.g. the Fig. 13 shows similarity between UO_2SeO_4 and UO_2SO_4 absorption spectra shapes. (In spectral region of highest absorption (380-440 nm), different measurements [5,68] of UO_2SO_4 differ from each other more than they differ from UO_2SeO_4 [36], for λ > 440 nm the S-Se difference can be explained by expanded measurement uncertainties as well).

[ii] With the respect to the standard euclidean norm (25), here without the N in denominator.

DOI: 10.1051/epjconf/201612802002

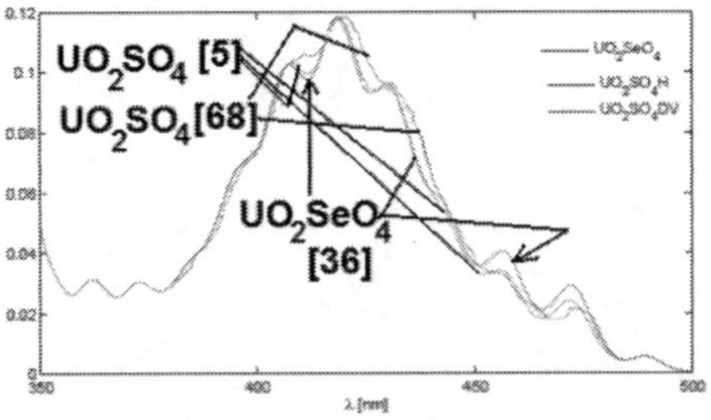

Fig. 13. Comparison between UO_2SeO_4 and UO_2SO_4 absorption spectra shows almost identical shapes.

Table 1: Uranyl monoselenate and uranyl biselenate thermodynamic properties for 25°C. The prime before β, i..e. 'β denote that in this study we subtracted log(7/5) from all stability constants (results from eventual $(UO_2)_2OH^{3+}$ neglection a posteriori), for uncorrected values see the first column in the Tab. 2.

	This study			study[19]		
log 'β_1°	2,48	±	0,20	2,38	±	0,06
log 'β_2°	3,11	±	0,10	3,70	±	0,15
ΔH_1° [kJ·mol^{-1}]	20,9	±	5,0	20	±	2
ΔH_2° [kJ·mol^{-1}]	19,2	±	6,3	31	±	4
ΔS_1° [J·mol^{-1}·K^{-1}]	126	±	21	123	±	6
ΔS_2° [J·mol^{-1}·K^{-1}]	119	±	17	180	±	15

Table 2: Uranyl monoselenate and uranyl bis(selenate) thermodynamic stability constants for 25°C.

	This study	[11]
log β_1°	2.63 ± 0.20	2.64 ± 0.01
log β_2°	3.26 ± 0.10	≤ 3.4

4.2 TRLFS spectroscopy

4.2.1. Uranyl–sulfate system for I=0.3 mol·dm^{-3}

The references in the Table 3 below (summarizing experimental stability constant values) stands for several chemometric methods [13,97,98] for data analysis used in J. Višňák Bachelor project – 1 = Factor analysis of the luminescence spectra of all solutions in series for the shortest delay between excitation laser pulse and ICCD camera signal accumulation t_1 = 150 ns (dentoted FASV ($t = t_1$)), 2 = Factor analysis of the time-resolved spectra of all solutions in series (monoexponential decay supposed, dependence of $\tau_{m,i}$ on i-th solution composition allowed, denoted FAČRSV in the thesis), 3 = Parallel factor analysis method (PARAFAC-3), NEA = The tabulated NEA value (the 18°C assigned value calculated from 25°C assigned NEA value plus Van't Hoff correction based on the ΔH° from PhreeqC PCHatches18.dat database)

Table 3. Concentrational stability constants β_n for the uranyl-sulfate complex species $[UO_2(SO_4)_n]^{2-2n}$ (n = 1, 2) for ionic strength I = 0.3 mol·dm^{-3} and temperature t = 25°C (recalculated from the 18°C measurement results by Van't Hoff equation with standard complexation enthalpies ΔH° from the PhreeqC database – PCHatches18.dat[24]).

ref.	log β_1			log β_2		
1	2.09	±	0.20	3.15	±	0.40
2	1.84	±	0.47	3.31	±	0.50
3	2.46	±	0.15	4.13	±	0.16
NEA	1.87	±	0.30	2.81	±	0.34

Experimental spectroscopic properties in temporal (Table 4) and wave-length/wave-number (Table 5, Fig. 14) domains are listed below.

Table 4. Luminescence life-times τ_m (τ_1 corresponds to UO_2^{2+}, τ_2 to UO_2SO_4 and τ_3 to $UO_2(SO_4)_2^{2-}$) for ionic strength I = 0.3 mold·dm^{-3} and measurement temperature 18°C in comparison with the literature data for 25°C.

ref.	τ_1 [μs]			τ_2 [μs]			τ_3 [μs]		
2	2.39	±	0,03	4.71	±	0.03	16.0	±	0.1
3	2.16	±	0,15	4.93	±	0.02	11.4	±	0.1
1	<3			4.7	±	0.3	11.5	±	0.3
1'	2.7	±	0,3	4.3			11.0	±	1.0

References in Table 4 are: 2 = FAČRSV [97,13] mean value (mean from $\tau_{m,i}$ over i), 3 = PARAFAC-3 value, 1 and 1' are tabulated literature data from other authors, 1 stands for I = 0.2 M, c = 0.05 M measurement and 1' for I = 1 M measurement. Since those two measurements doesn't differ much and our ionic strength lies between theirs, we can conclude that luminescence life-times presented in thesis [98] are in good agreement with literature data.

Table 5: Experimental luminescence spectroscopic properties of uranyl-sulfate system (I = 0.3 mol·dm^{-3}, t = 18°C, *pH* = 2).

species	$\omega_{gs,sym}$ [cm^{-1}]	$T_{0\text{-}0}$ [cm^{-1}]
UO_2^{2+}	875 ± 8	20 506 ± 21
UO_2SO_4	860 ± 8	20 363 ± 21
$UO_2(SO_4)_2^{2-}$	854 ± 8	20 274 ± 21

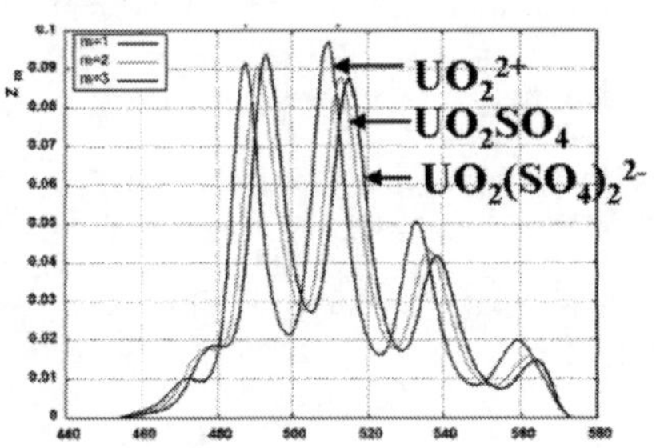

Fig. 14: Uranyl-sulfate system luminescence spectra

4.2.1. Uranyl–selenate system for I = 0.05 mol dm^{-3} and I = 1.1 mol·kg$_w^{-1}$

Uranyl – selenate system was investigated experimentally by TRLFS for I = 0.05 mol·dm^{-3} ionic strength within J. Višňák stay ad IPN Orsay under supervision of Dr. V.

Sladkov (Tab. 6[jj],7, [36]) and for higher ionic strength I = 1.1 mol·kg$_w^{-1}$ at Dep. Nucl. Chem., Prague (Fig. 15, for more results in this case [36]).

Table 6. Concentrational stability constants β_n for the uranyl-selenate complex species $[UO_2(SeO_4)_n]^{2-2n}$.

I	log β_1			log β_2		
0.05^{-0}	3.05	±	0.25			
1.1	1.54	±	0.45	1.75	±	0.50
1.1^{-0}	2.80	±	0.50	3.0	±	0.6
$[11]^{-0}$	2.64	±	0.01	≤ 3.4		
$[99]^{-0}$	2.38		0.06	3.70	±	0.15

Table 7. Experimental luminescence spectroscopic properties of uranyl-selenate system (I = 0.05 mol·dm^{-3}, t = 26°C, pH = 2).

species	$\omega_{gs,sym}$ [cm^{-1}]	$T_{0\text{-}0}$ [cm^{-1}]
UO_2^{2+}	870 ± 10	20 470 ± 20
UO_2SeO_4	850 ± 10	20 290 ± 20
$UO_2(SeO_4)_2^{2-}$	850 ± 10	20 200 ± 20

5 Computationals

Vast majority of calculations were done in Turbomole V6.5 program. Part of scalar relativistic effects were included via 60 electron Effective Core Potential (ECP) for uranium atom, all other electrons were, at this level of approximation treated as non-relativistic[kk]. The atomic basis sets used were def-SVP (denoted „S" in all tables in the section 6 below, [30,100]) and def-TZVPP (denoted „T", [29,30], for solvated models, for solvent water molecules only def-SVP atomic basis set was used to reduce the computational demands).

For the ground state properties, Hartree-Fock / Self-Consistent Field (SCF) method was used for comparison (as a method not including the electron correlation) and for generation of molecular orbital basis for post-HF wave-function based methods (MP2, CIS, CC) [81]. First method used was DFT with hybrid functional B3-LYP [28,82-85] and GGA functional B-P [82-87]. The latter in Turbomole V6.5 is implemented also within Resolution of Identity (RI-) approximation [90,91] which further save computational time allowing for larger supramolecular structures to be investigated. The RI-methodology was used in post-HF methods as well.

[jj] The values in rows with ionic strength with upper index "-0" correspond to thermodynamic ("activity defined") stability constants.
[kk] this excludes the possibility for luminescence life-time and μ_m computational estimation as they are beyond-scalar relativistic effects, but can be a good starting point for equilibrium geometry, vibrational frequency and hydration estimation

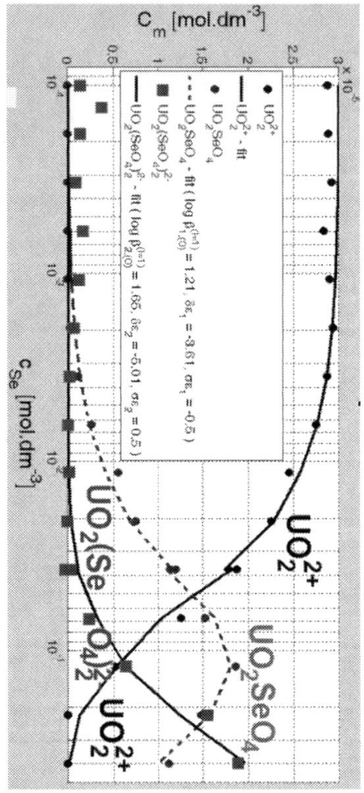

Fig. 15. Speciation diagram for the uranyl–selenate (UO_2^{2+}-SeO_4^{2-}-H_2O) system (I=1.1 mol·kgw^{-1}, pH=2) measured by TRLFS method. Full circles correspond to PCA predicted experimental molar concentrations of uranyl (UO_2^{2+}), uranyl monoselenate (UO_2SeO_4) and uranyl bis(selenate) ($UO_2(SeO_4)_2^{2-}$). The diagram has been exported as screenshot from Matlab [105] routine written by A. Vetešník from J. Višňák's algorithm.

For DFT calculations the D3 dispersion correction was applied [93]. Except few chosen cases which are denoted by subscript „$^{-D}$" in section 6 Tables.

The ref. 9 in Table 8 "4cDFT/2z" stands for all-electron four component Dirac equation-based DFT/B3-LYP within Dyall double-zeta atomic basis set [101] for bare UO_2^{2+} done in the DIRAC 13 program [88].

All figures have been exported from Molden [104] screenshots.

6 Quantum chemical calculation results

6.1 Uranyl, UO_2^{2+}

For the uncoordinated uranyl, we present results for the computed ground state properties for bare UO_2^{2+} group (Tab. 8), for the hydrated form - pentaaquauranyl complex $[UO_2(H_2O)_5]^{2+}$ (Table 9) and for certain explicit hydrated supramolecules $[UO_2(H_2O)_5]^{2+}$. n H_2O (Table 10). For the respective lowest lying excited triplet state properties, please see Tables 11-15.

The bare UO_2^{2+} SCF and DFT calculations were supplemented[ll] by 3-fragment counterpoise BSSE correction in the equilibrium geometry (single point BSSE calculation), where fragments were the individual atoms and were assigned by charge number +2 for uranium atom and 0 for oxygens[mm].

The results for the CC2 method are not presented, since, in agreement with previous studied done by others on, e.g. ozone molecule, this method is unsuitable for equilibrium geometry determination.

Table 8. Bare UO_2^{2+} ground state properties. The molecule is predicted to be linear by all methods below.

ref	method basis	ω_{sym} [cm^{-1}]	R_{min} [pm]	$E-E_{ref}$ [hartree]	ΔE_{BSSE} [10^{-3} hartree]
1	B3LYP/S	1049.7	169.6	0.17730	16.12
2	B3LYP/T	1044.1	169.3	0.00000	0.69
3	B-P/S	992.1	171.3	-0.19203	16.24
4	B-P/T	986.8	171.1	-0.36726	1.87
5	SCF/S	1243.3	164.4	2.45459	9.43
6	SCF/T	1240.3	163.9	2.26490	-0.04
7	MP2/S	938.2	172.8	1.63834	
8	MP2/T	943.9	171.5	1.31395	
9	4cDFT/2z		171.8		

Supposed linear geometry of central uranyl group was in $[UO_2(H_2O)_5]^{2+}$ slightly broken, in particular for the "C_{2v}-like" molecular models. The bond angle $_{yl}O-U-O_{yl}$ in the uranyl group will be denoted a in all tables below.

Table 9. Pentaaquauranyl $[UO_2(H_2O)_5]^{2+}$ in vacuum, ground state properties.

	model	method basis	ω_{sym} [cm^{-1}]	R_{min} [pm]	α [°]	$E-E_{ref}$ [hartree][nn]
1	C_{2v}	B3LYP/S	947.7	174.6	177	+102 cm^{-1}
2	D_{5h}	B3LYP/S	945.0	174.7	179	+0.6439
3	C_{2v}	B3LYP/T	946.6	174.1	177	+72 cm^{-1}
4	D_{5h}	B3LYP/T	946.9	174.1	180	0.0000
5	D_{5h}	B-P/S	888.6	176.9	176	+230 cm^{-1}
6	C_{2v}	B-P/S	893.8	176.7	176	+459 cm^{-1}
7	C_{2v}	RI-B-P/S	893.9	176.7	175	+212 cm^{-1}
8	D_{5h}	RI-B-P/S	888.5	176.9	176	+0.05843
9	C_{2v}	RI-B-P/T	892.8	176.1	176	+66 cm^{-1}
A	D_{5h}	RI-B-P/T	892.9	176.2	180	-0.576509
B	C_{2v}	SCF/S	1118	168.9	178	+140 cm^{-1}
C	D_{5h}	SCF/S	1117	169.0	180	+4.78537
D	C_{2v}	SCF/T	1112	168.3	178	+184 cm^{-1}
E	D_{5h}	SCF/T	1119	168.3	180	+4.13471
F	C_{2v}	MP2/S	887.3	177.3	177	+64 cm^{-1}
G	C_{2v}	MP2/T	888.1	175.6	177	
H	D_{5h}	MP2/S	887.1	177.3	180	+2.91340
I	D_{5h}	MP2/T		176.0	180	+0.91216

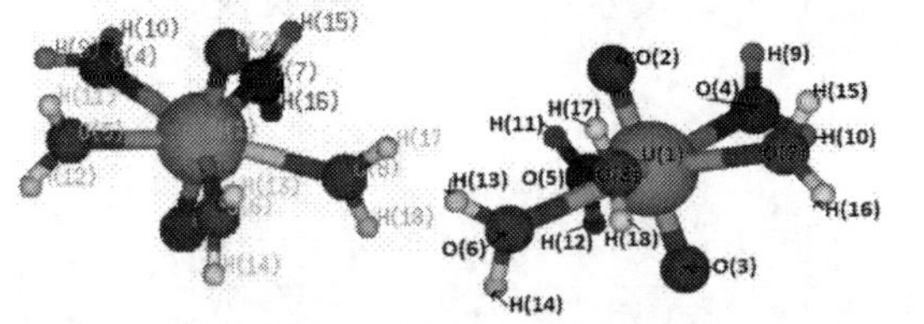

Fig. 16. DFT/B3LYP/defSVP calculations (left) The C_{2v}-like molecular geometry, please note the upper left coordinated water molecule (H(9)-O(4)-H(10)) which whole lies in the equatorial plane (plane perpendicular to the uranyl group) unlike the four others (which define planes perpendicular to the equatorial one). On right the D_{5h}-like molecular geometry (which is yet found global energy minimum) is depicted – the symmetry group is, in fact lower, supposed pentagon of donor oxygen atoms is deformed, but note that all water molecules are almost perpendicular to the equatorial plane.

The energy difference between the two structures in Figure 16 is between 64 cm^{-1} and 220 cm^{-1} depending on the chosen method (Table 9, rightmost column). In the most cases, the C_{2v}-like molecular geometries correspond to saddle point with one small imaginary frequency connected to tangential projection of rotation of the specific coordinated water (H(9)-O(4)-H(10)) from the left part of Figure 16 (results of geometry DFT/B3LYP optimization for the small atomic bais set def-SVP). The recalculation in larger basis set def-TZVPP let to the rotation of water molecules for the D_{5h} structure (Figure 17).

For the explicitly hydrated $[UO_2(H_2O)_5]^{2+}.nH_2O$ structures we started with model preserving the D_{5h} approximate symmetry (n = 10, left in Fig. 18).

The uranyl group vibrational modes for explicitly hydrated structures coupled to other supramolecular vibrational modes, as a result, Table 10 lists only one "typical frequency" and the other modes will be later discussed in the Supplementary Information.

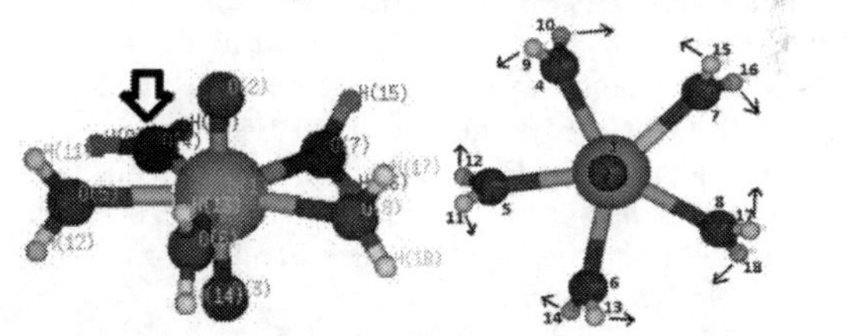

Fig. 17. DFT/B3LYP/defTZVPP equilibrium geometries (left): C_{2v}-like model with the water in equatorial coordination plane highlighted by an arrow, (right): top view on D_{5h}-like model showing slight rotation of water molecules with the respect to the $U-O_{water}$ axis – the water planes have dihedrals (like O(8)-U(1)-O(6)-H(13)) 68.5-68.9° instead of 90° (perpendicular do equatorial plane).

[ll] The BSSE corrections here and in the proceedings contribution below are not included in the listed electronic energies and their differences.
[mm] Since O(0)≡U(2+)≡O(0)) rounded partial charges are the same for a wide variety of quantum chemical methods and partial charge determining methodics.
[nn] $E_{ref} = E(D_{5h}$,B3LYP/T) = -1009.20134718 hartree for the first four rows. Since only computations for the same molecule within same or similar method are reasonably comparable, the E_{ref} respectively vary row by row. For clarity, the matching and particular E_{ref} values will be provided in the Supp. Inf. (values are in Hartree, for values in cm^{-1} the reference is the value for the same method and same basis given in hartree).

Table 10: Hydrated Pentaaquauranyl $[UO_2(H_2O)_5]^{2+}$. *n* H_2O, ground state properties.

	n	method basis	ω_{sym} [cm^{-1}]	R_{min} [pm]	α [°]	$E-E_{ref}$ [hartree]
1	10	B3LYP/S	888.7	177.5	178	1.6746
2	10	SCF/S	1074	170.5	179	9.5488
3	10	B3LYP/S	925.8	176.2	180	1.6810
4	10	SCF/S	1083	170.3	180	9.5509
5	10	B3LYP/S	869.7	177.2	179	1.6712
6	10	B3LYP/T	905.4	176.4	179	1.0832
7	10	B3LYP/T	927.3	175.5	180	1.0257
8	10	RI-B-P/S	837.9	179.7	178	0.6488
9	**10**	**RI-B-P/T**	**840.3**	**179.2**	**178**	**0.0000**
A	12	B3LYP/S	918.1	176.5	178	0.6567
B	12	B3LYP/S	909.9	176.4	179	0.6682
C	**12**	**B3LYP/T**	**910.9'**	**176.0**	**178**	**0.0000**
D	12	B3LYP/S	899.2	176.5	177	0.6575
E	12	B3LYP/T	905.0	176.9	177	0.0005
F	20	B3LYP/S	895.4	177.5	173	0.0062
G	20	B3LYP/S	890.9	177.6	173	0.0042
H	**20**	**B3LYP/S**	**876.6**	**178.1**	**179**	**0.0000**
I	20	B3LYP/S	908.5	176.6	179	0.1081
J	20	B3LYP/S	890.9	177.6	173	0.0042
K	20	SCF/S	1074	170.5	179	11.6142
L	20	B3LYP/T	893.7	177.1	173	-0.6724
M	55	RI-B-P/S	838.1'	176.6	172	-2.9190
N	**55**	**B3LYP/S**	**922.5'**	**175.6**	**176**	**0.0000**

A need to eliminate even the one small imaginary vibrational frequency (Fig. 18) let us to structure with opened rings of solvent water molecules (right in Fig. 18). To close the ring again, $[UO_2(H_2O)_5]^{2+}$. 12 H_2O models with either reflexion plane (Fig. 19, left) or inversion centre (Fig. 19, right) have been adopted. Figure 20 shows the largest structure investigated yet with 55 solvent water molecules.

The two fragment BSSE correction where one fragment are two coordinated water molecules (which will left the coordination sphere in possible reaction leading to $[UO_2(\kappa^2\text{-}SO_4)(H_2O)_3]$) for D_{5h}-like $[UO_2(H_2O)_5]^{2+}$ in vacuum yield 20.26 mH (53.20 kJ/mol) and 3.37 mH (8.84 kJ/mol) for def-SVP and def-TZVPP atomic basis sets for DFT single-point calculation at the ground electronic state equilibrium geometry.

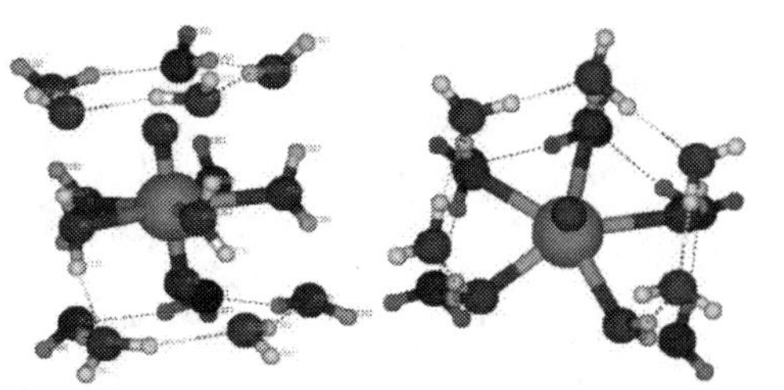

Fig. 18. $[UO_2(H_2O)_5]^{2+}\cdot 10H_2O$ DFT/B3LYP/T°° equilibrium geometries, left model (referenced as "6" in Tab. 10), resembling D_5 geometry corresponds to local stationary point with one imaginary vibrational frequency (ie 23.1 cm^{-1}). The right model („7" in Tab. 10) has 12.6×10^3 cm^{-1} higher energy than the left one and has lower symmetry, but corresponds to true local minimum.

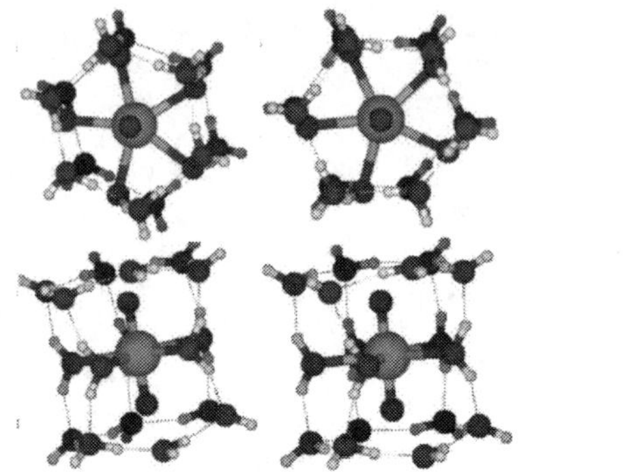

Fig. 19. $[UO_2(H_2O)_5]^{2+}\cdot 12H_2O$ DFT/B3LYP/T eq. geometries ("C" and "E" in Tab. 10). While the solute has D_{5h} symmetry, solvent molecules forms D_{6h} (left) or D_6 (right). The left structure is favoured by 99 cm^{-1}. The prime in frequency value in the respective row „E" stands for „frozen nuclei" approximation applied to solvent molecules.

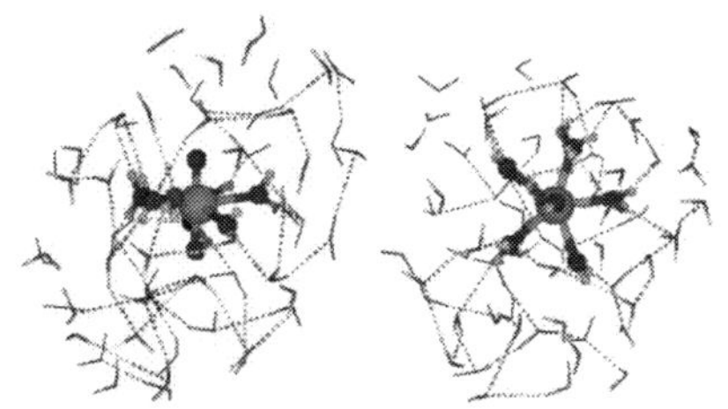

Fig. 20. $[UO_2(H_2O)_5]^{2+}.55H_2O$ equilibrium geometry (front and top view on supramolecule) for DFT/B3LYP/defSVP method. Initial geometry was generated by Packing [102] from in vacuum optimized pentaaquauranyl, positions for 55 solvent water molecules have been optimized by Turbomole default universal force-filed (solute fixed) and then only solute geometry was optimized by DFT.

For a bare uranyl in vacuum (Table 11), the ordering of the two lowest lying triplet states ($^3\Delta_g$, $^3\Phi_g$) is predicted to be $T_x(^3\Phi_g) < T_x(^3\Delta_g)$ by TDDFT while the Hartree-Fock(SCF)/CIS predicts the opposite ($T_x(^3\Phi_g) > T_x(^3\Delta_g)$). However, both methods predict the same ordering of symmetric mode vibrational frequencies $\omega_{es,sym}$ and electronic excitation induced U-O bond elongation ΔR.

Table 11. Excited state spectroscopic properties, bare uranyl, UO_2^{2+}, in vacuum. Excitation energies *T* are in 10^3 cm^{-1}

method basis	$\omega_{es,sym}$ [cm^{-1}]	ΔR [pm]	$T_{0,0}$	$T_{v,ex}$	$T_{v,d}$	state
B3LYP/S	903.9	6.2	14.6	16.8	13.3	$^3\Phi_g$
	927.9	4.6	17.3	18.7	16.7	$^3\Delta_g$
B3LYP/T	903.2	6.2	14.8	17.0	13.6	$^3\Phi_g$
	925.0	2.6	17.2	18.6	16.6	$^3\Delta_g$
CIS-HF/S	1111	2.6	20.6	21.4	20.5	$^3\Delta_g$
	1088	3.8	25.6	26.8	25.0	$^3\Phi_g$
CIS-HF/T	1103	2.8	20.0	23.8	19.7	$^3\Delta_g$
	1085	3.8	25.2	27.8	24.5	$^3\Phi_g$

For the pentaaquauranyl, $[UO_2(H_2O)_5]^{2+}$, in vacuum, only one lowest lying excited triplet state has been studied. Unlike for the ground state, for the excited state, C_{2v} equilibrium geometry is preferred by 177 cm^{-1} (389 cm^{-1})

°° def-TZVPP AO basis was used for the central complex $[UO_2(H_2O)_5]^{2+}$ and def-SVP AO basis for the 10 solvent water molecules.

over the D_{5h} one due to the TDDFT/defSVP (defTZVPP). However, according to CIS-HF/defSVP the D_{5h} excited state minimum is by 36 cm^{-1} lower. In the table 12 the column "Model" shows molecular geometry of the lower and upper electronic states considered and for the excited state also the respective irreducible representation ($D_{5h}(1e_2')$ stands for an approximate symmetry, the true symmetry is $C_2(1a)$ – see Fig. 17 (right)), "55" stands for $D_{5h} \rightarrow D_{5h}(1e_2')$, "22" stands for $C_{2v} \rightarrow C_{2v}(1b_1)$ and "52" stands for $D_{5h} \rightarrow C_{2v}(1b_1)$.

Table 12: Excited state spectroscopic properties, Pentaaquauranyl in vacuum[pp].

ref	method basis	$T_{0,0}$ [cm^{-1}]	$T_{v,ex}$ [cm^{-1}]	$T_{v,d}$ [cm^{-1}]	model
1	B3LYP/S	19 193	**20 497**	18 459	55
2		19 064	19 993	**18 306**	22
3		19 284	**20 497**	**18 306**	52
4	B3LYP/T		20 264	18 465	55
5			19 803	18 115	22
6	CIS-HF/S	24 254	**24 951**	24 130	55
7		24 276	24 847	**24 038**	22
8		24 516	**24 951**	**24 038**	52
9	CIS-HF/T		24 770		55
10			24 673	23 785	22
11	CIS(D)/S		19132		55
12	CIS(D)/T		18373		22
13	CIS(D)/S		18922		55
14	CIS(D)/T		18835		22

Table 13. Excited state spectroscopic properties, Pentaaquauranyl in vacuum.

	model	method basis	$\omega_{es,sym}$ [cm^{-1}]	ΔR [pm]	α_{es} [°]
1	55	B3LYP/S	833.1	4.58	180
2	22		834.5	4.63	176
3	55	B3LYP/T		4.88	180
4	22			4.67	176
5	55	CIS-HF /S	994.3	2.62	180
6	22		997.3	2.62	178
7	22	CIS-HF/T		2.77	178

Table 14. Excited state spectroscopic properties, hydrated Pentaaquauranyl $[UO_2(H_2O)_5]^{2+}$. *n* H_2O.

	n	method basis	$\omega_{es,sym}$ [cm^{-1}]	ΔR [pm]	α_{es} [°]	$T_{0,0}$ [cm^{-1}]
1	10	B3LYP/S-D	771.8	4.70	178	19966
2	10	CIS/S	960.4	2.52	179	24819
3	10	B3LYP/S	840.1	4.51	180	20181
4	10	B3LYP/T		4.48	180	
7	12	B3LYP/S	803.3	4.60	177	20264
8	12	B3LYP/T	808.1	4.60	177	
9	20	B3LYP/S-D	802.8	4.64	179	19789

Table 15. Excitation energies, hydrated Pentaaquauranyl $[UO_2(H_2O)_5]^{2+}$. *n* H_2O.

	n	method basis	T_e [cm^{-1}]	T_{de} [cm^{-1}]	T_a [cm^{-1}]
1	10	B3LYP/S-D	20969	19382	20217
2	10	CIS/S	25322	24605	24986
3	10	B3LYP/S	21119	19617	20431
4	10	B3LYP/T	20964	19411	20224
5	10	B3LYP/S	21130	19585	20408
6	10	B3LYP/S		19311	20147
7	12	B3LYP/S	21081	19520	20348
8	12	B3LYP/T	20846	19309	20116
9	20	B3LYP/S-D	20792	19032	19956

[pp] For CIS(D), the ground state was calculated by RI-MP2 method.

6.2 Uranyl Monosulfate, UO_2SO_4

Uranyl Monosulfate complex, UO_2SO_4, as mentioned in 2.3 section has several isomeric forms differing by the uranyl group coordination number (5 or 6, Fig. 21) and dentacity of coordinated sulfate ligand (1 (Fig. 22) or 2 (Fig. 21)).

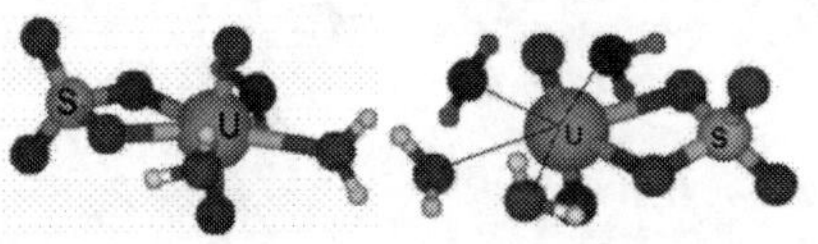

Fig. 21: $[UO_2(\kappa^2\text{-}SO_4)(H_2O)_3]$ (κ^2co5, left) and $[UO_2(\kappa^2\text{-}SO_4)(H_2O)_4]$ (κ^2co6, right) ground state equilibrium geometries by DFT/B3-LYP/defTZVPP method

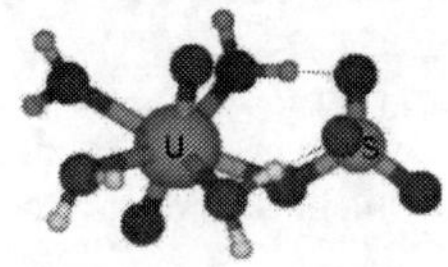

Fig. 22: $[UO_2(\kappa^1\text{-}SO_4)(H_2O)_4]$ (κ^1co5) ground state equilibrium geometry by DFT/B3-LYP/defSVP method

For the fourth possibility, $[UO_2(\kappa^1\text{-}SO_4)(H_2O)_5]$ (κ^1co6) no energy minimum has been found up to now. All geometry optimizations led either to water molecule drive out from the coordination plane or to hydrogen transfer from coordinated water molecule to sulphate ligand creating more stable $[UO_2(\kappa^1\text{-}HSO_4)(OH)(H_2O)_4]$. Ground state properties in vacuum are summarized in Table 16, κ^2co5h and κ^1co5h in first column stands for $[UO_2(\kappa^2\text{-}SO_4)(H_2O)_3]\cdot H_2O$ and $[UO_2(\kappa^1\text{-}SO_4)(H_2O)_4]\cdot H_2O$ respectively.

Table 16. Uranyl Monosulphate in vacuum, ground state properties. E_{ref} = -1555,953902886 H (hartree) for the first four rows (the κ^2co5 isomer), E_{ref} = -1632,415734279 H for others.

isomer	method basis	$\omega_{sym,1}$ [cm^{-1}]	$\omega_{sym,2}$ [cm^{-1}]	R_{min} [pm]	α [°]	E-E_{ref} [H]
κ^2co5	DFT/S	841.7	886.0	177.8	168	1.0036
κ^2co5	DFT/T	851.2	881.2	177.5	167	0.0000
κ^2co5	SCF/S	1005	1057	171.3	171	6.4363
κ^2co5	SCF/T	1024	1059	170.8	171	5.4066
κ^2co6	DFT/S	861.0	889.4	177.6	171	1.1099
κ^2co6	DFT/S	859.3	890.0	177.6	171	1.1112
κ^2co5h	DFT/S	849.3	884.0	177.9	169	1.0986
κ^2co5h	DFT/T	858.7	883.6	177.4	169	0.0000
κ^2co6	DFT/T	868.0	890.0	177.1	170	0.0125
κ^2co6	SCF/S	1024	1061	171.0	174	6.9158
κ^1co5h	DFT/S	892.6		177.5	173	
κ^1co5	DFT/S	886.7		177.7	167	1.1133

The water molecule drives out from the coordination plane can be exploited for preliminary energy comparison between κ^2co6 and κ^2co5 isomers (at the DFT/B3LYP/defTZVPP level, κ^2co5 is favoured by 12.5

 DOI: 10.1051/epjconf/201612802002

mH = 2736 cm^{-1} = 32.7 kJ/mol, inclusion of ZPE difference, 33.8 kJ/mol). The standard reaction gibbs function change ΔG° for the chemical reaction

$$\left[UO_2\left(\kappa^2-SO_4\right)\left(H_2O\right)_3\right]\cdot H_2O$$

$$\left[UO_2\left(\kappa^2-SO_4\right)\left(H_2O\right)_4\right]$$

can be estimated as $\Delta G°$ = 31.6 kJ/mol, corresponding to $K°$ = $2.9\cdot10^{-6}$ for standard temperature T = 298.15 K (i.e. the κ^2co5 is significantly preferred over κ^2co6 in vacuum). Similarly, for the reaction

$$\left[UO_2\left(\kappa^2-SO_4\right)\left(H_2O\right)_4\right] \qquad \left[UO_2\left(\kappa^1-SO_4\right)\left(H_2O\right)_4\right]$$

$\Delta G°$ = 13.8 kJ/mol and $K°$ = $3.85\cdot10^{-3}$ (T = 298.15 K) (i.e. the κ^1- isomer is present in less than 0.4% in equilibrium within this level of theory). DFT functional in the Table 16 was B3LYP.

The coordination of sulphate ligand led to uranyl group symmetric vibrational mode coupling with the ligand vibrational modes. The result are two modes separated by 33^{+22}_{-11} cm^{-1} (the sub/superscript corresponds to extremal deviations) in average. The red-shifted part of the pair corresponds to the synchronous move of z-axis (uranyl group) projections of uranyl oxygen atoms and non-donor oxygen atoms of the sulphate and is denoted as $\omega_{sym,1}$ in Table 16. The blue-shifted part corresponds to the anti-synchronous change of the z-coordinate ($\omega_{sym,2}$ in Tab. 16). This effect has been smaller for the excited electronic state and only one frequency is listed in the respective table of excited state properties (Table 18).

Table 17: Excited state spectroscopic properties, Uranyl Monosulphate in vacuum.

isomer	method basis	T_e [cm^{-1}]	isomer	method basis	T_e [cm^{-1}]
κ^2co5	B3LYP/S	17640	κ^2co6	B3LYP/S	18231
κ^2co5	B3LYP/T	17412	κ^2co5h	B3LYP/S	18418
κ^2co5	CIS-HF/S	24537	κ^2co6	CIS-HF/S	24477
κ^2co5	CIS-HF/T	24349	κ^1co5h	B3LYP/S	18866
κ^2co6	B3LYP/S	18011			

Table 18. Excited state spectroscopic properties, Uranyl Monosulphate in vacuum. The method is CIS-HF for all three calculations, basis is denoted "b." in the second column.

isomer	b.	$\omega_{es,sym}$ [cm^{-1}]	ΔR [pm]	α_{es} [°]	$T_{v,d}$ [cm^{-1}]	T_{00} [cm^{-1}]
κ^2co5	S	942.7	2.39	171	23931	24064
κ^2co5	T		2.53	170	23671	
κ^2co6	S	944.5	2.46	173	23838	23963

Properties of the lowest triplet state are summarized in Tables 17 and 18. The basis set change from defSVP ("S") to defTZVPP ("T") led to decrease of 1.03 H in κ^2co5 excited state energy (PES minima both) resulting in change of the adiabatic excitation energy T_a from 24248 cm^{-1} to 24024 cm^{-1}

The two fragment BSSE correction was computed for the single point in the DFT/B3-LYP equilibrium geometries for the κ^2co5 isomer. The sulphate ligand with charge -2 was considered as one of the fragments. The resulting corrections were 60.87 mH (defSVP) and 12.15 mH (defTZVPP).

The hydration effect will be shown for κ^2co5 isomer here only (it is the most stable isomer in vacuum, however, the hydration can change the isomeric relative stabilities significantly). The other isomers hydration will be discussed in the Supplementary Information. We have used the RI-DFT method for the 55 solvent water structures to reduce the computational demands[99], but are aware that the compatible GGA functional B-P underestimates vibrational frequencies (which are still to be assigned to the corresponding modes) and is in general less accurate than the hybrid B3-LYP functional. The small energy differences for n = 10 (Tab. 19, first three rows) are rather results of similar input geometries for the geometry optimization. More elaborate study, where configuration space will be rigorously searched by MD should be done. The respective equilibrium geometries are shown in Fig. 23-25.

Table 19. Hydrated $[UO_2(\kappa^2\text{-}SO_4)(H_2O)_3]\cdot nH_2O$, ground state properties and vertical excitation energy T_e (n = 10: DFT/B3LYP/defSVP method, n = 55: RI-DFT/B-P/defSVP method)

n	$\omega_{sym,1}$ [cm^{-1}]	$\omega_{sym,2}$ [cm^{-1}]	R_{min} [pm]	α [°]	T_e [cm^{-1}]	$E\text{-}E_{ref}$ [mH]
10	904.7	914.0	176.6	176	20406	0.00
10	842.8	855.0	179.0	174	20282	6.21
10	862.5	881.7	177.9	178	20327	4.30
55			177.8	168		0.00
55			179.2	167		423.00

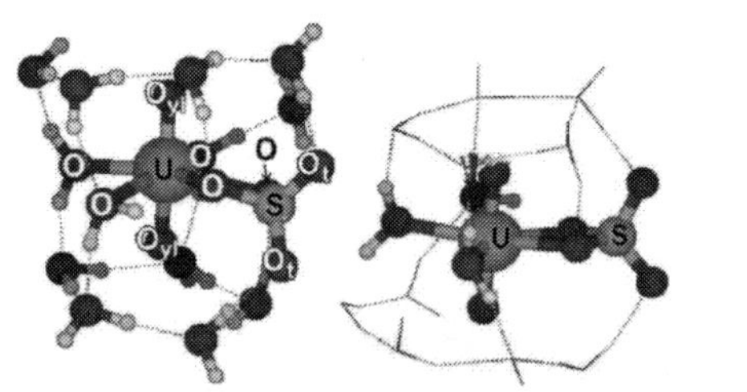

Fig. 23. Equilibrium geometries computed by DFT/B3LYP/defSVP method for $[UO_2(\kappa^2\text{-}SO_4)(H_2O)_3]\cdot10H_2O$ model. The left side corresponds to first, the right side to third row in Table 19.

[99] Furthermore, for n = 55 solvating water molecules models here, only the coordinates of solute atoms have been varied within the DFT energy optimization. The solvent atomic coordinates have been optimized only once prior the DFT calculation by universal force-field calculation (the Turbomole default UFF [103], first row under thick line in Table 19) or used from the Packing program [102].

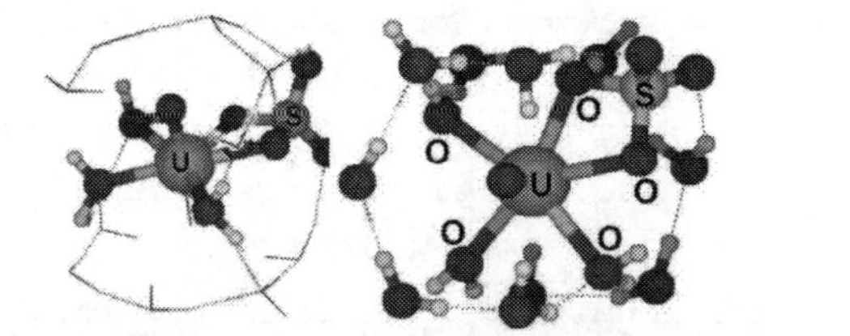

Fig. 24. Equilibrium geometry computed by DFT/B3LYP/defSVP method for $[UO_2(\kappa^2\text{-}SO_4)(H_2O)_3]\cdot 10nH_2O$ model (second row in Tab. 19). The right side is a top view on the molecule on left side, donor oxygens are labeled for clarity.

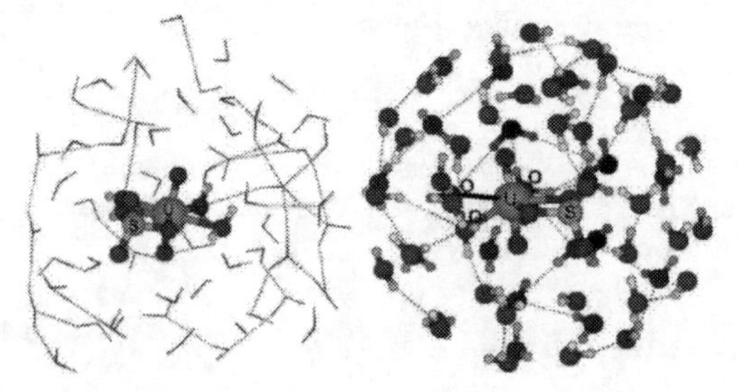

Fig. 25. Equilibrium geometries computed by RI-DFT/B-PB3LYP/defSVP method for $[UO_2(\kappa^2\text{-}SO_4)(H_2O)_3]\cdot 55H_2O$ model. The left side corresponds to first, the right side to second respective row (under the thick line) in Table 19.

From a batch of 8 possible initial geometries based on idea of "sandwich" structure, where the equatorial coordination plane of the complex is the "filler" and two pentagonal ring of water molecules (connected by H-bonds to each other and to donor oxygen atoms of the complex) are "bread" (the initial geometries different by whether the rings have the asymmetric H-bonds oriented in the same or in the opposite directions, by dihedral angles, etc.) several local PES minima were found (the three lowest are presented in Table 19).

In order to investigate possible mechanisms of peak broadening, we have used a luminescence peak model

$$V(\nu) = \sum_{k=1}^{8} B(\nu_{\max,k},T)\cdot \exp\left(-\frac{(\nu-\nu_{\max,k})^2}{2\sigma^2}\right), \quad (63)$$

where ν is wave-number, $\nu_{max,k}$ corresponds to the vertical excitation energy (the spread of vertical excitation energies T_e with the respect to different solvation surroundings should be similar to the spread of T_{00} values) of the k-th local PES minima discussed in previous paragraph, σ is an empirical width-parameter (s = 60 cm^{-1} was used here) and T is thermodynamic temperature (for demonstrability we used triple of the standard temperature here),

$$B(\nu_{\max,k},T) = \frac{1}{Z}\exp\left(-\frac{h\nu_{\max,k}}{k_BT}\right), \quad (64),$$

is a Boltzmann factor, Z is partial sum,

$$Z = \sum_{k=1}^{8} \exp\left(-\frac{h\nu_{\max,k}}{k_BT}\right), \quad (65)$$

h, k_B are Planck and Boltzmann constants. The Figure 26 shows that resulting peak FWHM is around 180 cm^{-1} (the total spread of T_e (highlighted by dashed vertical lines in Fig. 26) within over batch of initial geometries is 680 cm^{-1}). The experimental luminescence 0' → 0 peak FWHM is around 700 cm^{-1}. The discrepancy is a result of very simplified model of one kind of peak broadening only.

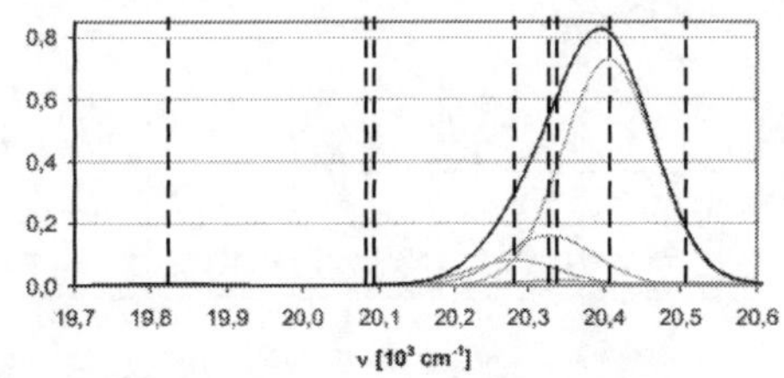

Fig. 26. Peak modelled according to eq. (63) for luminescence peak-width explanatory discussion.

6.3 Uranyl bis(sulfate), $UO_2(SO_4)_2^{2-}$

We briefly present vacuum results for $UO_2(SO_4)_2^{2-}$ in Tab. 20, the method DFT-D3/B3-LYP was used, basis is denoted "b.", ω_{sym} in cm^{-1} stands for simple average over all modes where symmetric stretching of uranyl group was significant. For the corresponding equilibrium geometries, see Fig. 27.

For κ^{22}co5 and κ^{12}co5 isomers, TDDFT/B3-LYP excited state geometry optimization led to $T_{0\text{-}0}$, $\omega_{es,sym}$, ΔR and α_{es} values in Tab. 21.

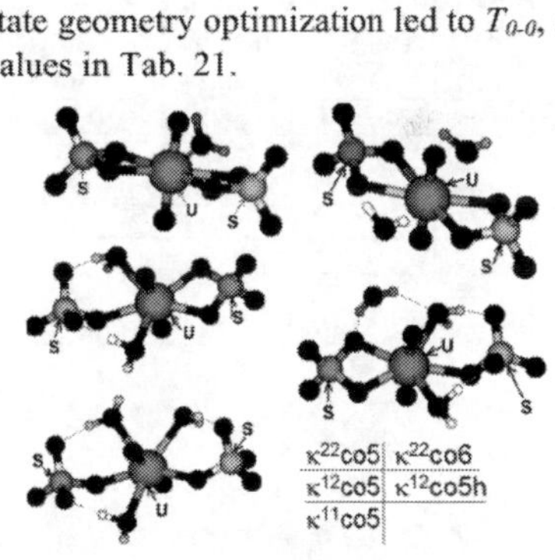

Fig. 27. Equilibrium geometries for several uranyl bis(sulphate) isomers, DFT-D3/B3LYP/def-TZVPP method.

Table 20. Uranyl Bis(sulphate) ($UO_2(SO_4)_2^{2-}$) in vacuum, ground state properties and vertical excitation energies. $E_{ref,1}$ = -2102,2549546 H for the first two rows (the κ^{22}co5 isomer $[UO_2(SO_4)_2(H_2O)]^{2-}$), $E_{ref,2}$ = -2178,7093575 H for κ^{12}co5 and κ^{22}co6 isomers $[UO_2(SO_4)_2(H_2O)_2]^{2-}$ and $E_{ref,3}$ = -2255,1593788 H for κ^{11}co5 ($[UO_2(SO_4)_2(H_2O)_3]^{2-}$) and κ^{12}co5h ($[UO_2(SO_4)_2(H_2O)_2]^{2-}$. H_2O). Uranyl Tris(sulphate) in vacuum (last two rows, E_{ref} = -2724,5375210 H)

isomer	b.	ω_{sym} [cm^{-1}]	T_e [cm^{-1}]	R_{min} [pm]	α [°]	E-E_{ref} [H]
κ^{22}co5	S	859	19468	179.0	176	1.3732

κ^{22}co5	**T**	**880**	**19341**	**178.4**	**176**	**0.0000**
κ^{12}co5	S	865	19762	178.2	177	1.4678
κ^{12}co5	**T**	**847**	**19762**	**177.8**	**177**	**0.0000**
κ^{11}co5	S	839	19622	178.4	177	1.5663
κ^{11}co5	**T**	**853**	**19336**	**177.8**	**177**	**0.0000**
κ^{22}co6	S	872	19942	178.7	179	1.4756
κ^{22}co6	**T**	**877**	**19792**	**178.1**	**179**	**0.0044**
κ^{12}co5h	S	872	19410	178.5	178	1.5678
Trissulf	S	834	20646	179.1	180	1.8642
Trissulf	**T**	**836**	**20468**	**178.3**	**180**	**0.0000**

Table 21: Excited state spectroscopic properties, Uranyl Bis(sulfate) ($UO_2(SO_4)_2^{2-}$) in vacuum.

isomer	b.	$\omega_{ek,sym}$ [cm^{-1}]	ΔR [pm]	α_{es} [°]	$T_{0,0}$ [cm^{-1}]
κ^{22}co5	S	762.7	3.62	175	18749
κ^{22}co5	T		3.66	175	
κ^{12}co5	S	820.2	1.44	176	13082

6.4 Uranyl tris(sulphate), $UO_2(SO_4)_3^{4-}$

For uranyl tris(sulphate), $[UO_2(\kappa^2\text{-}SO_4)_3]^{4-}$, inclusion of one incomplete (30 water molecules) hydration sphere on vibrational frequency wsym (increased by 45 cm^{-1}) and vertical excitation energy (T_e, decreased by 418 cm^{-1}, Tab. 20 – last two rows, Tab. 22 – first row). Equilibrium geometry for the lowest local PES minimum yet found is depcited in Fig. 27.

Table 22: $[UO_2(\kappa^2\text{-}SO_4)_3]^{4-}\cdot 30H_2O$ (E_{ref} = -5013,7054062 H), R_{min} = 177.1 pm and α = 180° for all lines below.

ref.	ω_{sym} [cm^{-1}]	T_e [cm^{-1}]	E-E$_{ref}$ [mH]
1	879.2	20228	0
2	878.0	20233	0.41
3	879.2	20225	0.77
4	879.4	20221	0.81
5	876.7	20233	1.18
6	878.8	20231	1.20
7	884.2	20224	1.29
8	880.5	20514	8.13

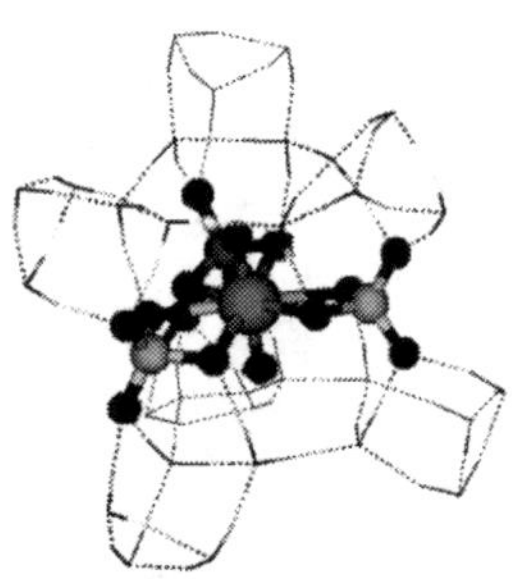

Fig. 27: $[UO_2(\kappa^2\text{-}SO_4)_3]^{4-}\cdot 30H_2O$ equilibrium geometry, DFT-D3/B3LYP/def-SVP method.

6.5 Uranyl mono(selenate), UO_2SeO_4

Table 23. Uranyl Monoselenate (UO_2SeO_4) in vacuum. ground state properties. E_{ref} = -1555.953902886 H (hartree) for the first four rows (the κ^2co5 isomer). E_{ref} = -1632.415734279 H for others.

isomer	b.	ω_{sym} [cm^{-1}]	T_e [cm^{-1}]	R_{min} [pm]	α [°]	E-E$_{ref}$ [H]
κ^2co5	S	883	18406	177.8	168	1.1505
κ^2co5	**T**	877	18250	177.4	167	0.0000
κ^2co5h	S	880	18772	177.9	170	1.2325
κ^2co6	S	885	18720	177.5	170	1.2482
κ^1co5	S	885	19409	177.7	167	1.2441
κ^1co5	**T**	882		178.3	167	0.0000

6.6 Other complex molecules

We greatly apologize to the reader, quantum chemical calculations of the 1:2 and 1:3 sulphates as well as selenates and carbonates will be introduced in the Supplementary Information and further performed in Višňák dissertation thesis.

7 Conclusions

For UO_2^{2+} - SO_4^{2-} and UO_2^{2+} - SeO_4^{2-} systems, good agreement between here presented experimental spectroscopic and thermodynamic properties and data of other authors has been found. The uranyl – selenate complex species are slightly less stable than the corresponding uranyl – sulphate species (Tab. 2, Tab. 3) since selenate is weaker Lewis base than sulphate while uranyl is strong Lewis acid [9,36]. The shapes of absorption spectra are identical for uranyl sulphates and selenates within measurement error, the luminescence spectral shapes are very close to each other as well. This can be demonstrated on similar spectroscopic properties $T_{0\text{-}0}$ and ω_{gs} (Tab. 5, Tab. 7 – however, please note the different ionic strengths).

For the reference, the $[UO_2(H_2O)_5]^{2+}$ 0'→0 deexcitation energy has been experimentally determined as $T_{0\text{-}0}$ = (20490±30) cm^{-1} (Tab. 5 and Tab. 7 average[rr]) and the effective uranyl symmetric stretching vibrational frequency (in ground electronic state) ω_{gs}=(873±13) cm^{-1}. The one ligand coordination leads to bathochromic shift[ss] of 140±30 cm^{-1} (0.7%) for the sulphate (Tab. 5) and of 180±30 cm^{-1} for the selenate. The two ligand coordination produce bathochromic shift 230 and 270 cm^{-1} for sulphate and selenate respectively. The vibrational frequency is decreased by ≈ 20 cm^{-1} by any coordination, however, this effect is within the measurement error. Yet only for hydrated uranyl luminescence spectrum, the ΔR parameter fit has been done, ΔR = (16±4) pm (Fig. 6).

Quasi-relativistic, DFT-D3/B3-LYP/def-SVP theoretical estimates of spectroscopic parameters have been done for complex species in vacuum and in aqueous solution (modelled by explicit solvation, yet, by small number of surrounding water molecules). For $[UO_2(H_2O)_5]^{2+}$ the vacuum $T_{0\text{-}0}$ prediction is 19.8x10^3 cm^{-1} (Tab. 12) and the $T_{0\text{-}0}$ vary between (19.8-20.3)x10^3 cm^{-1} (Tab. 14) for solvated model. The latter value is in good agreement with the experimental value, in particular, if we

[rr] FWHM *not* included in the uncertainity.

[ss] The $T_{0\text{-}0}$ decrease.

include the FWHM into experimental uncertainty. The $[UO_2(H_2O)_5]^{2+}$ in vacuum vibrational frequency ω_{gs} has been predicted as 947 cm^{-1} by DFT/B3-LYP and as 888 cm^{-1} by MP2 (Tab. 9). Inclusion of 10, 12 and 20 solvation water molecules led to gradual wgs decrease to 927, 911 and 894 cm^{-1} respectively (Tab. 10, see the energy lowest "B3LYP/T" model). The values for 55 water molecules are less reliable since approximative structure optimization has been used in that cases. The $[UO_2(H_2O)_5]^{2+}.20H_2O$ ω_{gs} of 894 cm^{-1} seems to be very promisive in comparison with experimental data.

The theoretical predictions for ΔR are less than 5 pm, while experiment suggest more than 12 pm (average 16 pm). The discrepancy could be explained by simplicity of the one-mode-oscillator model for ΔR fit from the experimental data. In is likely, that more vibrational modes with similar (or similar to multiples) frequencies will resolve the luminescence spectra and the 16 pm therefore corresponds to sum of equilibrium position shifts over all vibrational modes coupling to the luminescence transition. The more detailed Franck-Condon factors analysis should clarify this.

Facing technical problems, we were able to determine $T_{0\text{-}0}$ (within TDDFT/B3-LYP/def-SVP) only for uranyl bis (sulphate) κ^{22}co5 isomer in vacuum as 18749 cm^{-1} (Tab. 21) which accounts for 535 cm^{-1} (Tab. 12) bathochromic shift, 230% of the experimental value (please note, the vacuum theoretical estimate and, of course, hydrated experimental values are compared). The κ^{12}co5 prediction is even more deviated.

As a substitute, we can discuss the decrease of the vertical excitation energy T_e with the respect to (sulphate, selenate) ligand coordination–then the predicted bathochromic shift of T_e is at least 560 cm^{-1} (for 10 solvating water molecules, Tab. 15, Tab. 19), as compared to 140 cm^{-1} for $T_{0\text{-}0}$ experimentally determined change. From Tab. 15 and Tab. 22 we can also deduce the bathochromic shift (i.e. lowering) of T_e in comparison $[UO_2(\kappa^2\text{-}SO_4)_3]^{4-}\cdot 30H_2O$ and $[UO_2(H_2O)_5]^{2+}\cdot 20H_2O$. Te of the former is by 327 cm^{-1} lower.

Explanation of the predicted unrealistically high bathochromic shifts might be that sulphate coordinated species, in particular those with negative net charge, have loosely bounded LUMO electrons. Better chosen atomic basis set should be used to describe them.

The vibrational frequency of κ^2co5 uranyl monosulphate isomer ($[UO_2(\kappa^2\text{-}SO_4)(H_2O)_3]\cdot n\ H_2O$) with $n = 10$ water molecule, 909 cm^{-1} (Tab. 19), is by 18 cm^{-1} lower than the corresponding quantity for $[UO_2(H_2O)_5]^{2+}$ (927 cm^{-1}, Tab. 10).

For 30 water molecules surrounded uranyl tris (sulphate), the vibrational frequency has been determined as 879 cm^{-1} (Tab. 22), lower by 15 cm^{-1} than ω_{gs} for $[UO_2(H_2O)_5]^{2+}\cdot 20H_2O$ (Tab.10).

For uranyl – selenate system, even greater difficulties have been met. With predicted Te for bis(selenate) being greater than for monoselenate) and similar anomalies. The detailed report will be given in the Supplementary Information.

CIS (with HF as ground-state reference) method has been shown as unfeasible for spectroscopic properties predictions. CIS(D), on the other side, might be interesting for comparison (and we will aim to apply CC-family methods as well) (Tab. 12).

The peak full-width-half-maxima (FWHM) can be explained and modelled by $T_{0\text{-}0}$ and ω_{gs}/Franck-Condon factor calculations based on (Maxwell-Boltzmann-factor-weighted average of) different PES local minima, isomers and solvation surroundings. Of course, another potential source of band-width is theoretical possibility for transitions from another electronic excited states quasi-degenerate with the lowest lying one (see the chapter 2.1 and the Supplementary Information[tt])

Important would be to estimate spectral norms, mm and therefore absorption and luminescence full, not normalized, spectra. For this purpose, however, spin-orbital interaction is essential. In best case, the four-component Dirac-Coulomb-Gaunt Hamiltonian and wave-functions should be used. The electronic transition probability between the excited $\Omega = 1_g$ state, which is superposition of several spin-multiplicity components (let us denote the corresponding coefficients as $c(1_g, 2s+1)$),

$$\left|1_g\right\rangle = \sum_{s=0}^{s_{max}} c\left(1_g, 2s+1\right)\left|{}^{2s+1}Y_{es}\right\rangle, \qquad (66)$$

$$\left|0_g^+\right\rangle = \sum_{s=0}^{s_{max}} c\left(0_g^+, 2s+1\right)\left|{}^{2s+1}Y_{gs}\right\rangle, \qquad (67)$$

similarly for the ground state $\Omega = 0_g^{\ +}$. We suppose that the ground state is dominated by the singlet component and excited state by a triplet component. The radiational transition probability $1_g \rightarrow 0_g^{\ +}$ is then given by formula

$$P\left(1_g \rightarrow 0_g^+\right) = \sum_{s=0}^{s_{max}} \left|c\left(0_g^+, 2s+1\right)\right|^2 \left|c\left(1_g, 2s+1\right)\right|^2 \left|\left\langle {}^{2s+1}Y_{es}\right|\hat{T}\left|{}^{2s+1}Y_{gs}\right\rangle\right|^2, \qquad (68)$$

where terms in the summation corresponds to singlet-singlet, triplet-triplet, quintuplet-quintuplet, ... transitions between hypothetical eigenstates of total molecular spin. The matrix elements correspond to $\Delta L_z = \pm 1$ dipole transitions.

We hope that the contribution presented can serve as a good methodological base for further, more accurate, theoretical predictions of uranyl (VI) complex species spectroscopic and thermodynamic properties.

Acknowledgements

I would like to thank the Department of Nuclear Chemistry for support. the Czech Student Grant Agency (grant no. SGS16/250/OHK4/3T/14) for material support. Special thanks are for experimental collaborators – Bc. Jakub Kuba. Dr. Vladimir Sladkov and Dr. Aleš Vetešník. Useful

---

[tt] Which will be placed on the first author's Academia.edu personal page in next weeks.

help and hints came from Dr. Šebesta, Fibich and Sláma. I would like to thank TESNAT 2016 Organizing Committee for hospitality and the opportunity to present my work. Access to computing and storage facilities: National Grid Infrastructure MetaCentrum (www.metacentrum.cz). under program CESNET LM2015042 is greatly appreciated.

References

1. R.J. Lakowicz, *Principles of Fluorescence Spectroscopy*, Springer, (2006).
2. G. Meinrath, Review Focusing on Aspects of Environmental Chemistry, (1998).
3. http://www.world-nuclear.org
4. W.G. Wright, J. Environ. Qual., **28**, No 4 (1999).
5. D. Vopálka, et al., J. Radioanal. Nucl. Chem., **286**: 681 (2010).
6. A. Vetešník, et al., Uranium, Mining and Hydrogeolology. Springer, p. 623-630. (2008).
7. J. Višňák, et al., The New Uranium Mining Boom, Edts: B. Merkel and M. Schipek, 643-652 (2012).
8. T. Vercouter, P. Vitorge, B. Amekraz, Ch. Moulin, Inorg. Chem., **47**, 2180-2189 (2008).
9. V. Sladkov, B. Fourest, F. Mercier, Dalton Trans., 7734-7740 (2009).
10. P. Lubal, J. Havel, Chem. Papers **51** (4) 213-220 (1997).
11. P. Lubal, J. Havel, Talanta **44**, 457-466 (1997).
12. M. Maeder, Y.-M. Neuhold, Science and Technology, Edt: S. Rutan and B. Walczak, **26**, Elsevier, (2007).
13. J. Višňák, Master Thesis, FNSPE, Czech Technical University (2010) (In Czech).
14. A. de Juan, R. Tauler, Anal. Chim. Acta, **500** (1–2): 195–210 (2003).
15. R. Bro, et al., Applied Spectroscopy Reviews. 32 (3): 237–261 (1997).
16. Ch. F. Beckmann, S.M. Smith, Tensorial Extensions of Independent Component Analysis for Multi-Subject FMRI Analysis, FMRIB, Technical Report TR04CB1, http://www.fmrib.ox.ac.uk.
17. R. Harshman, M. Lundy, Research methods for multimode data analysis, 122-215. Praeger, (1984).
18. R. Harshman, UCLA Working Papers in Phonetics, 16, 84 pp. (1970).
19. G.H. Golub, C. Reinsch, Numerische Mathematik. **14** (5): 403–420, (1970).
20. P. Atkins, J. De Paula, *Physical Chemistry*, W.H. Freeman and Company. p. 212 (2006).
21. K.G. Dyall, K. Jr. Faegri, *Introduction to Relativistic Quantum Chemistry*, Oxford Uni. Press, (2007).
22. P. Hohenberg, W. Kohn, Phys. Rev. **136** (3B): B864–B871, (1964).
23. W. Kohn, L.J. Sham, Phys. Rev. **140** (4A): A1133-A1138 (1965).
24. A. Szabo, N.S. Ostlund, Modern Quantum Chemistry: Introduction to Advanced Electronic Structure Theory, McGraw-Hill, (1989).
25. E. Runge, E.K.U. Gross, Phys. Rev. Lett. **52** (12): 997–1000 (1984).
26. M. Petersilka, U.J. Gossmann; E.K.U. Gross, Phys. Rev. Lett. **76** (8): 1212–1215 (1996).
27. C. Ullrich, *Time-Dependent Density-Functional Theory: Concepts and Applications*, Oxford University, (2012).
28. A.D. Becke, J. Chem. Phys. **98** (7): 5648–5652 (1993).
29. F. Weigend et al, Chem. Phys. Letters **294**, 143 (1998).
30. TURBOMOLE V6.5 2013, University of Karlsruhe and Forschungszentrum Karlsruhe GmbH, 1989-2007, http://www.turbomole.com.
31. F. Furche, et al., WIREs Comput Mol Sci, 4:91–100 (2014).
32. X. Cao, M. Dolg, J. Molec. Struct., 673, 203-209 (2004).
33. M. Dolg, in Proceedings, J. Grotendorst (Ed.), John von Neumann Institute for Computing, **3**, pp. 507-540 (2000).
34. V.E. Jackson, et al., J. Phys. Chem. A, 112, 4095-4099, (2008).
35. Z. Wang, et al., Environ. Sci. Technol., **38**, 5591-5597 (2004).
36. J. Višňák, J et al., In Lectures of colloquium on Radioanalytical methods, **IAA 14**, (2014).
37. J. T. Bell, R.E. Biggers, *J. Molecular Spectroscopy*, **25**, 312 (1986).
38. L.H. Jones, Spectrochim. Acta, **10**, 395-403, (1958).
39. J.J. Olivero, R.L. Longbothum, J. of Quantitative Spectroscopy and Radiative Transfer. **17** (2): 233–236 (1977).
40. G.K. Wertheim, et al., Review of Scientific Instruments. 45 (11): 1369–1371 (1974).
41. M. Gal, P.L. Goggin, J. Mink, Spectrochim. Acta **48A** (1992).
42. L.C. Andrews, *Special Functions of Mathematics for Engineers*, Oxford Science Publications, (1998).
43. W.H. Greene, *Econometric Analysis*, Prentice-Hall, (1993).
44. M. Born, R. Oppenheimer, Ann. Phys. **84**, 457 (1927).
45. J. Franck, Transactions of the Faraday Society, **21**: 536–542, (1926).
46. E. Condon, Phys Rev, **28**: 1182–1201, (1926).
47. A.S. Coolidge, H.M. James, R.D. Present, J. Chem. Phys, **4**: 193–211, (1936).
48. P.W. Atkins, R.S. Friedman, *Molecular Quantum Mechanics*, Oxford University Press, (1999).
49. V.I. Minkin, Pure and Applied Chemistry. **71**, 10, Pages 1919–1981.[uu]
50. E. Rudberg, Diploma work in Physical Chemistry, Royal Institute of Technology, Sweden (2003).
51. M. Abramowitz, I. A. Stegun, eds. "Chapter 22". *Handbook of Mathematical Functions with Formulas, Graphs, and Mathematical Tables*. Applied Mathematics Series. 55 Dover Publications. (1983).
52. https://en.wikipedia.org/wiki/Hermite_polynomials

[uu] The formula on the respective page 1941 there contains, however error/typo, the correct one should read

$$V = \frac{1}{2}\sum_{i,j=1}^{3N} \frac{\partial^2 V}{\partial q_i \partial q_j} q_i q_j \quad (66)$$

53. J. Višňák routine in Wolfram Mathematica [54].
54. Wolfram Research, Inc., Mathematica, Version 10.4, Champaign, IL (2016).
55. J.D.J. Ingle, S.R. Crouch, *Spectrochemical Analysis*, Prentice Hall, (1988).
56. J.-L. Chang, J. Mol. Spectrosc., **232**, 102-104, (2005).
57. R. Islampour, et al., J. Mol. Spectrosc. **194**, 179–184 (1999).
58. P.-A. Malmqvist, N. Forsberg, Chem. Phys., **228**, 227–240, (1998).
59. K. Pierloot K., E. van Besien, J Chem Phys **123**, 204309 (2005).
60. J. Mehra, H. Rechenberg, *The Historical Development of Quantum Theory*, Springer, 2001.
61. G. Budylin, et al., *Optics Express*, **21**, No. 18, (2013).
62. H. D. Burrows, S. J. Formosinho, *J. Chem. Educ.*, **55** (2), p 125 (1978).
63. Laser-Induced Spectroscopy, Helmholtz Zentrum Dresden Rossendorf website, http://www.hzdr.de.
64. H.T. Evans Jr., *Science*, **141**, 154-158, (1963).
65. R.G. Denning, Gmelin Handbuch der Anorganischen Chemie, Uran (Supplement) **A6**, 31-79, (1983).
66. R.G. Denning, et al., *Mol. Phys.*, **37**, 1089 - 1107, (1979).
67. B. Zhu, S.A. Pennell, D.K. Ryan, *Microchem. J.*, **115**, 51-57, (2014).
68. C. Hennig, et al., *AIP Conf. Proc.*, **882**, 262 (2007).
69. J. Šebera et al, *In Scientific report of the project for SÚRAO, the Czech Radioactive Waste Repository Authority for the final control date (Phase 4)*, FNSPE, Czech Technical University, p 27-42, (July 2009).
70. J.N. Brønsted, *J. Am. Chem. Soc.*, **44**, 877–898 and 938–948 (1922).
71. G. Scatchard, *Chem. Rev.*, **19**, 309–327 (1936).
72. E.A. Guggenheim, *Applications of Statistical Mechanics*, Clarendon Press, (1966).
73. M.R. Wright, *An Introduction to Aqueous Electrolyte Solutions*, section 10.6.15, Wiley, (2007).
74. I. Grenthe, H. Wanner, E. Osthols, *TDB-2, Guidelines for the extrapolation to zero ionic strength*, OECD Nuclear Energy Agency, (2000).
75. M. Bühl, et al., *Inorg. Chem.*, **50**, 299–308 (2011).
76. M. Zhou, L. Andrews, N. Ismail, C. Marsden, J. *Phys. Chem. A*, **104**, 5495-5502 (2000).
77. W.A. Gezahegne, et al., *Environ. Sci. Technol.*, **46**, 2228–2233 (2012).
78. B. Li, et al., *Comp. Theor. Chem.*, **1051**, 151–160, (2015).
79. L. Gagliardi, et al., *J. Phys. Chem. A*, **105**, 10602-10606, (2001).
80. P.D. Dau, et al., *J. Chem. Phys.*, **137**, 064315 (2012).
81. M. Head-Gordon, et al., *Chem. Phys. Lett.* **219**, 21 (1994).
82. P.A.M. Dirac, *Proc. Royal Soc.* (London) A, **123**, 714, (1929).
83. J.C. Slater, *Phys. Rev.*, **81**, 385, (1951).
84. S.H. Vosko, L. Wilk, M. Nusair, *Can. J. Phys.*, **58** 1200, (1980).
85. C. Lee, W. Yang, R.G. Parr, *Phys. Rev. B*, **37**, 785 (1988).
86. A.D. Becke, *Phys. Rev.* A **38**, 3098 (1988).
87. J.P. Perdew, *Phys. Rev.* B **33**, 8822 (1986).
88. DIRAC, a relativistic ab initio electronic structure program, Release DIRAC13 (2013), written by L. Visscher, et al.
89. A. Schäfer, et al., *Phys. Chem. Chem. Phys.*, **2**, 2187-2193, (2000).
90. K. Eichkorn, et al., *Chem. Phys. Letters* **242**, 652 (1995).
91. K. Eichkorn, et al., *Theor. Chem. Acc.* **97**, 119 (1997).
92. M. Sierka, A. Hogekamp and R. Ahlrichs; *J. Chem. Phys.* **118**, 9136, (2003).
93. S. Grimme, J. Antony, S. Ehrlich, Krieg, *J. Chem. Phys*, **132**, 154104 (2010).
94. E.B. Wilson, Jr., *J. Chem. Phys.* **9** (1): 76–84 (1941).
95. A. Malijevksý, University of Chemistry and Technology, Prague, 2009. http://old.vscht.cz.
96. J. Višňák, et al., *Annual report of the Department of Nuclear Chemistry 2011-2012*, Prague. http://www.jaderna-chemie.cz/data.
97. J. Višňák, et al., *Annual report of the Department of Nuclear Chemistry 2008-2010*, Prague. http://www.jaderna-chemie.cz.
98. J. Višňák, Laserem indukovaná fluorescenční spektroskopie s časovým rozlišením jako nástroj pro přímou speciaci uranu o nízkých koncentracích, Bachelor project, FNSPE, Czech Technical University in Prague (2008).
99. V. Sladkov, *J. Chromatography A*, **1276**, 120-125, (2013).
100. A. Schäfer, *J. Chem. Phys.*, **97**, 2571 (1992).
101. A.S.P. Gomes, K.G. Dyall, L. Visscher, *Theor Chem Acc*, **127**: 369, (2010).
102. J.M. Martinez, L. Martinez, *J. Comput. Chem.*, **24**:819-825,(2003).
103. A.K. Rappé, et al., *J. Am. Chem. Soc.*, **114**(25), 10024–10035, (1992).
104. G.Schaftenaar and J.H. Noordik, *J. Comput.-Aided Mol. Design*, **14**, 123-134, (2000).
105. MATLAB and Statistics Toolbox Release 2012b, The MathWorks, United States.
106. B. Drobot, et al., Chem. Sci., **6**, 964-972, (2015).

Radionuclide concentration in cabbage samples due to gamma radiation in Samsun, Turkey

Aydan Altıkulaç [1], *Hasan* Gümüş [2]

[1]Muğla Sıtkı Koçman University, Ula Ali Koçman MYO, Department of Electricity and Energy, Muğla, Turkey
[2] Ondokuzmayıs University, Faculty of Sciences and Art, Department of Physics, Samsun, Turkey

Abstract. Establishing of radioactivity planes in foodstuff has emphasis because it allows the evaluation of population exposure to radiation by take nourishment. In this paper, the activity concentrations of ^{226}Ra, ^{232}Th, ^{40}K and ^{137}Cs were determined in cabbage samples collected from Samsun city of Turkey using a gamma ray spectrometry method with a HPGe detector. The mean concentration value of ^{226}Ra, ^{232}Th,^{40}K and ^{137}Cs in cabbage samples were 1.11±0.03 Bqkg^{-1}, 1.44±0.04 Bqkg^{-1}, 743.75±21.21 Bqkg^{-1} and 0.18±0.003 Bqkg^{-1}, respectively. The calculated total annual effective dose received from ^{226}Ra, ^{232}Th, ^{40}K and ^{137}Cs due to cabbage samples by population of Samsun province was quite lower than the World average value as suggested by UNSCEAR.

1 Introduction

The sources of radioactivity in the environment have natural, terrestrial and anthropogenic origins which are resulted from nuclear trial and the processing of nuclear power stations. It is known that air residue of artificial radionuclides be formed as a result of the significant series of nuclear gun tests and nuclear disaster such as the Chernobyl disaster in 1986. The primary resource of artificial radionuclides in Turkey is the nuclear contamination from the Chernobyl accidents. During the process of nuclear waste, the airbone particles may be intercepted by plants or return to help soil. Therefore, plants may get radioactivity nuclides by deposition of radioactive waste on the foodstuff directly and by absorption from the soil Measurement of the concentrations of radionuclides present in foodstuffs permits the determine of the dose caused by the intake of foodstuff [1]. Since gamma radiation provides information about excess lifetime cancer risk, determining gamma dose rate has also importance [2]. The aim of this study is to determine the specific activities of radionuclides such as ^{226}Ra, ^{232}Th, ^{40}K and ^{137}Cs in cabbage samples grown in the Samsun city which is so close to the nuclear central that is going to be built in Turkey and calculate the annual effective dose.

2 Materials and methods

The samples were put inside on the detector and take counted for a term of 80.000 s. The net area under the specific peaks in energy spectrum was calculated by taking counts owing to Compton scattering of suitable peaks. From the net area of a clear peak, the activity concentrations in the samples were acquired from the following equation.

$$C\,(Bqkg^{-1}) = \frac{Cn}{\varepsilon\ P_{\gamma}\,Ms} \qquad (1)$$

Where C is the activity concentration of the radionuclide in sample given in Bqkg^{-1}, C_n is the count ratio under the specific peak, ε is the detector efficiency at the certain γ-ray energy, P_{γ} is the exact progression probability of the specific γ- ray and M_s is the mass of the sample [4]. The energy and efficiency calibration of the detector were performed using photopeaks from radioactive standards IAEA- RGU-1, IAEA-RGTh-1 and IAEA-RGK-1 reference materials [5]. Calibration sources were prepared with the same geometry as the vegetable samples thus didn't need to a geometry correction. The efficiency tables were formed by using known activities of photopeaks of intense gamma emitters from uranium series nuclides, ^{214}Bi and ^{214}Pb, thorium series nuclides ^{228}Ac and ^{208}Tl and gamma transitions of 1461 keV for K were also used [6]. Relative efficiency at 1332 keV ^{60}Co is %35. The ^{226}Ra activity of the samples was detected owing to the density of 351.9 keV and 609.3 keV gamma marks of ^{214}Pb and ^{214}Bi separately.

3 Result and discussion

3.1 Activity concentrations

The activity concentrations in cabbage samples ranged from 0.38±0.01 to 2.7±0.08 with an average 1.11±0.03 Bqkg^{-1} for ^{226}Ra; from 0.5±0.01 to 2.55±0.08 with an average 1.44±0.04 Bqkg^{-1} for ^{232}Th; from 0.1±0.003 to 0.71±0.02 with an average 0.18±0.003 for ^{137}Cs. The activity concentrations of ^{137}Cs are quite low values in many samples. As regarding of ^{40}K, Table.1 presents its measured activity concentrations in cabbage samples. The activity concentrations of ^{40}K varied between 373.37 and 1164.87±36.25 with an average 743.75±21.21. It can be said that the studied radionuclides (^{226}Ra, ^{232}Th, ^{40}K and ^{137}Cs) were defined in all cabbage samples.

Table 1. Activity concentrations of ^{226}Ra, ^{232}Th, ^{40}K and ^{137}Cs in evaluated in cabbage samples

Sample ID	Activity concentrations of cabbage samples ^{226}Ra	^{232}Th	^{40}K	^{137}Cs
1	0.75±0.02	1.98±0.05	753.37±24.87	0.25±0.008
2	0.62±0.02	0.7±0.02	866.75±27.37	0.16±0.005
3	1.6±0.05	1.16±0.03	1035.5±32.12	0.22±0.007
4	0.38±0.01	0.63±0.02	940.12±28.5	0.12±0.003
5	0.47±0.01	1.25±0.03	339.62±11.87	0.11±0.003
6	0.26±0.008	0.57±0.01	399.37±12.5	0.1±0.003
7	1.07±0.03	2.55±0.08	1221.5±39.5	0.27±0.007
8	0.66±0.02	2.05±0.06	1164.87±36.25	0.71±0.02
9	1.52±0.05	6.02±0.19	915.25±29.62	0.22±0.007
10	0.7±0.02	0.5±0.01	373.37±11.72	0.36±0.01
11	2.7±0.08	1.72±0.05	650.25±21.25	0.20±0.007
12	0.71±0.02	1.9±0.06	883±28.12	0.25±0.007
13	1.35±0.04	0.98±0.02	440.5±14.12	0.13±0.003
14	1.48±0.04	0.91±0.02	430.37±14.37	0.21±0.007
Average	1.11±0.03	1.44±0.04	743.75±21.21	0.18±0.003

3.2 Dose Estimation

Annual effective dose from estimated to evaluate the threat of radiation for human fitness. Annual effective dose is calculated by

$$H_{T,r}=\sum(U_iC_r^i)g_{T,r} \quad (2)$$

where i is foodstuff group, U^i and C_r^i are annual consumption ratio (kg) and radionuclide activity concentration (Bqkg^{-1}), respectively for their factor, $g_{T,r}$ is dose transformation factor for r radionuclide (SvBq^{-1}). Dose transformation factor of ^{226}Ra, ^{232}Th, ^{40}K and ^{137}Cs radionuclides for the adult members of society are 4.5x10^{-8}, 2.3x10^{-7}, 6.2x10^{-9} and 1.3x10^{-8} SvBq^{-1}, respectively [7,8,9]. The annual effective dose due to each radionuclide is shown in Table 2. It can be seen from Table 2 that ^{40}K conduced the maximum to the annual dose generated by intake of foodstuff. The sum of ^{40}K doses ensured by foodstuff adds to mean value of 33.66±0.5 μSvy^{-1}, which shows the mean annual effective dose for the Turkish consumer.

Table 2. Dose coefficients and appointed effective dose values

Radioisotopes	Activity intake (Bq)	AED (μSvBq-^{1})	Appointed effective dose (μSvy-^{1}) Range	Average
^{226}Ra	8.10±0.21	0.045	0.08±0.01-0.52±0.12	0.36±0.08
^{232}Th	10.51±0.29	0.23	1.02±0.3-12.30±1.14	2.94±0.61
^{40}K	5.429±147.53	6.2 x 10^{-3}	15.37±0.53-55.28±1.78	33.66±0.5
^{137}Cs	1.31±0.02	1.3 x 10^{-2}	0.002±0.01-0.71±0.26	0.01±0.003

The calculated total annual effective dose (AED) received from ^{137}Cs vary from 0.002±0.01 to 0.71±0.26 with a mean of 0.01±0.003 μSvy^{-1}. This paper based on that humans living in Samsun province be exposed radiation dose nearly 38 μSvy^{-1} from cabbage consumption is quite lower than the World average value (290 μSvy^{-1}) that established by UNSCEAR [10].

4 Conclusion

It was observed that the mean annual effective dose due to ^{40}K quite higher than ^{137}Cs. The total annual effective dose due to consumption of foodstuff was less than the average world value that announced as 290 μSvy^{-1} by UNSCEAR. As a result, the study showed that there is no radiological risk for the people living in Samsun province.

References

1. R. Keser, K.F. Görür, N. Akçay, *J.Sci Food Agric,* **91**, 987-991(2011).
2. Ş.Turhan, K. Abdullah, A. Varınlıoglu, *Isotopes in Environmental and Health Studies,* **43**, 249-256 (2007).
3. D. Mascanzoni, J. Radioanal. Nucl. Chem., 249, 245-249 (2001).
4. S. Yoshida, Y. Muramatsu, J. *Environ. Radioact.,* **41,** 183-205 (1998).
5. C. Canbazoğlu, M. Dogru, *J. Radioanal Nucl. Chem.* **295**,1245-1249 (2013).
6. A. Parmaksız, Y. Ağuş, *Radiaiton Effects and Defects in Solids,* (2014).
7. C. Kuwahara, et al., *Sci. Total Environ.,* **345**, 165-173 (2005).
8. G. Kirchner, O. Daillant, *Sci. Total Environ.,* **222**, 63-70 (1998).
9. M.I. Gasso, et al., *Radiat. Protect. Dosim.,* **87**, 213-216 (2000).

Contributions of each isotope in structural material on radiation damage in a hybrid reactor

Mehtap Günay

İnönü University, Science and Art Faculty, Physics Department, Malatya, Turkey

Abstract. In this study, the fluids were used in the liquid first-wall, blanket and shield zones of the designed hybrid reactor system. In this study, salt-heavy metal mixtures consisting of 93–85% $Li_{20}Sn_{80}$ + 5% SFG-PuO_2 and 2-10% UO_2, 93–85% $Li_{20}Sn_{80}$ + 5% SFG-PuO_2 and 2-10% NpO_2, and 93–85% $Li_{20}Sn_{80}$ + 5% SFG-PuO_2 and 2-10% UCO were used as fluids. In this study, the effect on the radiation damage of spent fuel-grade (SFG)-PuO_2, UO_2, NpO_2 and UCO contents was investigated in the structural material of a designed fusion–fission hybrid reactor system. In the designed hybrid reactor system were investigated the effect on the radiation damage of the selected fluid according to each isotopes of structural material in the structural material for 30 full power years (FPYs). Three-dimensional analyses were performed using the most recent MCNPX-2.7.0 Monte Carlo radiation transport code and the ENDF/B-VII.0 nuclear data library.

1 Introduction

Hybrid reactor system has been developed where fusion and fission occur at the same time. Hybrid reactor has an enormous potential to provide a clean, cheap and unlimited energy source. Important advantage of hybrid reactor system is that it has a subcritical reactor system in which reactor can operate securely [1-6].

A substantial amount of spent fuel grade (SFG) plutonium (Pu) from current nuclear reactors has been stored for future use. The isotopic distribution of spent fuel-grade plutonium is 2.4% ^{238}Pu, 58.5% ^{239}Pu, 24% ^{240}Pu, 11.2% ^{241}Pu, and 3.9% ^{242}Pu [7]. In this study, $Li_{20}Sn_{80}$ molten salt was used, which has low melting temperatures and low vapour pressure, to acquire sufficient tritium breeding. Beryllium (Be) is used as a neutron multiplier [8]. In this study, a Be zone with a thickness of 3 cm was used between the liquid first-wall and the blanket. In this study, 9Cr2WVTa ferritic steel was chosen as a structural material because it has low activation, which can help to extend the lifetime of the reactor system. The main objective of this study is to investigate the effect of the selected fluid on the radiation damage according to each isotopes of structural material in the structural material for the designed hybrid system.

2 Method

2.1 Numerical calculation and radiation damage

In Table 1 is shown the radial build of the designed hybrid reactor system. Calculation of all the parameters of the fission and fusion reactors, accelerator-driven systems, and other areas of nuclear technology depend on the cross-section data. The experimental data are limited for neutron-produced reactions in certain energy ranges. The data of the ENDF/B-VII libraries are very important for theoretical calculations [9-13]. This study was performed with neutron wall loadings of 10 MW/m^2 and fusion power of 4000 MW.

Table 1. The radial build of the hybrid reactor system design

Inboard side		Outboard side	
Zone	r (cm)	Zone	r (cm)
SS316LN	276	Plasma	667
Vacuum vessel[a]	278	SOL	695
SS316LN	294	Liquid First Wall[d]	697
GAP	296	Be[e]	700
Shield[b]	301	Blanket[d]	750
Ferritic Steel[c]	350	Ferritic Steel[c]	754
Blanket[d]	354	Shield[b]	804
Be[e]	404	GAP	838
Liquid First Wall[d]	407	SS316LN	840
SOL	409	Vacuum vessel[a]	866
Plasma	437	SS316LN	868

[a] 80% SS316LN, 20% H_2O
[b] 60% 9Cr2WVTa, 40% (93–85% $Li_{20}Sn_{80}$ + 5% SFG-PuO_2 and 2-10% UO_2, 93–85% $Li_{20}Sn_{80}$ + 5% SFG-PuO_2 and 2-10% NpO_2, and 93–85% $Li_{20}Sn_{80}$ + 5% SFG-PuO_2 and 2-10% UCO)
[c] 100% 9Cr2WVTa
[d] 93–85% $Li_{20}Sn_{80}$ + 5% SFG-PuO_2 and 2-10% UO_2, 93–85% $Li_{20}Sn_{80}$ + 5% SFG-PuO_2 and 2-10% NpO_2, and 93–85% $Li_{20}Sn_{80}$ + 5% SFG-PuO_2 and 2-10% UCO
[e] 100% Be

Structural materials are exposed to high neutron flux, γ-rays, and energetic particles that continuously eject atoms from their lattice sites. This leads to various types of damage in the material due to neutron attenuation such as hardening, swelling and embrittlement, which directly influence its function. The abilities to reduce the activation and radiation damage to structural materials are among the most important advantages of liquid walls, particularly the "thick" liquid wall concept in the fusion reactor. Radiation effects are the consequence of radiation damage on the mechanical and physical properties of the solid. Whereas the hydrogen atom and its isotopes produced by (n, p), (n, d), and (n, t) reactions can diffuse out of the first wall material with high temperatures, the helium atoms produced by (n, α) and (n, He -3) reactions will accumulate in the first wall of the hybrid reactor and produce helium gas bubbles. These reactions will limit the operational lifetime of the structural material of the reactor [14-17].

3 Numerical Results

In this study, the fluids were composed of molten salt 93-85% $Li_{20}Sn_{80}$ + 5% SFG-PuO_2 as the main constituent with increasing mole fractions in 2–10% rates of the heavy metals UO_2, NpO_2, and UCO. The effect of UO_2, NpO_2, UCO on the radiation damage parameters were investigated according to each isotopes of structural material.

Fig. 1 and Fig. 2 shows comes from Fe isotope of the biggest contribution and from C isotope of the smallest contribution at the total proton and deuterium production rate, respectively, in the structural material for used fluids. In Fig. 1, contribution at the total proton production rate in the structural material of Fe isotope for used fluids was obtained about $5x10^5$, $1x10^3$, 10, $7x10^2$, $5x10^4$, $4x10^3$ times higher than C, V, Cr, Mn-55, Ta-181, W isotopes, respectively. In Fig. 2, contribution at the total deuterium production rate in the structural material of Fe isotope for used fluids was obtained about $1x10^4$, $3x10^2$, $2x10^2$, $2x10^2$ times higher than C, V, Cr, Mn-55 isotopes, respectively. In Fig. 1 and Fig. 2, the contributions to the total proton and deuterium production rate of Fe isotope in the structural material was found as approximately 93% and 99%, respectively.

Fig. 3 shows comes from Cr isotope of the biggest contribution and from V isotope of the smallest contribution at the total tritium production rate in the structural material for used fluids. In Fig. 3, contribution at the total tritium production rate in the structural material of Cr isotope for used fluids was obtained about $1x10^2$, 20 and 4 times higher than V, Mn-55 and Fe isotopes, respectively. In Fig. 3, the contributions to the total tritium production rate of Cr isotope in the structural material was found as approximately 77%.

Fig. 4 shows comes from Fe isotope of the biggest contribution and from V isotope of the smallest contribution at the total He-3 production rate in the structural material for used fluids. In Fig. 4, contribution at the total He-3 production rate in the structural material of Fe isotope for used fluids was obtained about $2x10^5$, 6 and 10 times higher than V, Cr and Mn-55 isotopes, respectively. In Fig. 4, the contributions to the total He-3 production rate of Fe isotope in the structural material was found as approximately 79%.

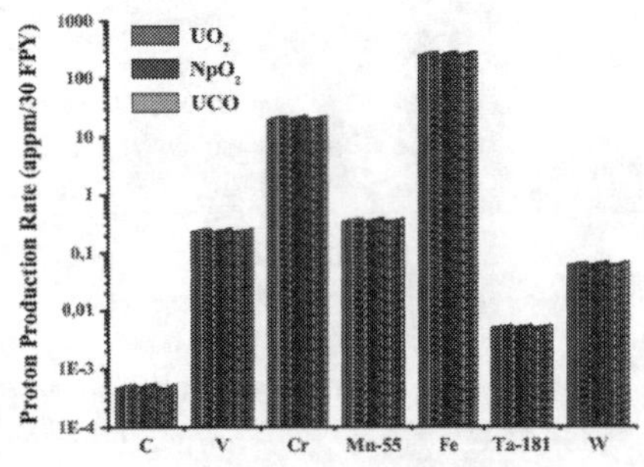

Fig. 1. The contribution of each isotope in the structural material to the total proton production rate for the mixture components of UO_2, NpO_2 and UCO (2%, 4%, 6%, 8%, 10%) in the system.

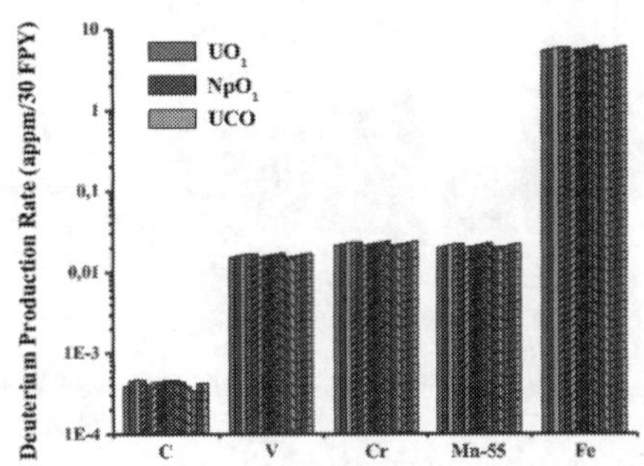

Fig. 2. The contribution of each isotope in the structural material to the total deuterium production rate for the mixture components of UO_2, NpO_2 and UCO (2%, 4%, 6%, 8%, 10%) in the system.

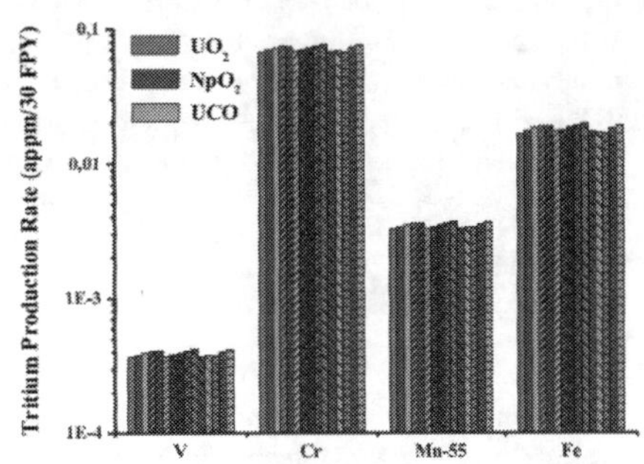

Fig. 3. The contribution of each isotope in the structural material to the total tritium production rate for the mixture components of UO_2, NpO_2 and UCO (2%, 4%, 6%, 8%, 10%) in the system.

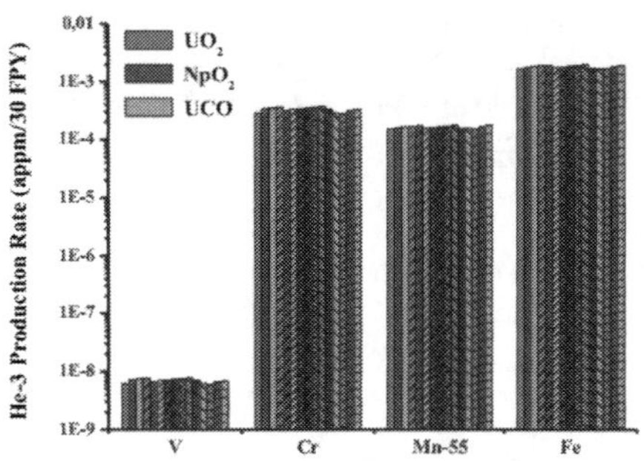

Fig. 4. The contribution of each isotope in the structural material to the total He-3 production rate for the mixture components of UO_2, NpO_2 and UCO (2%, 4%, 6%, 8%, 10%) in the system.

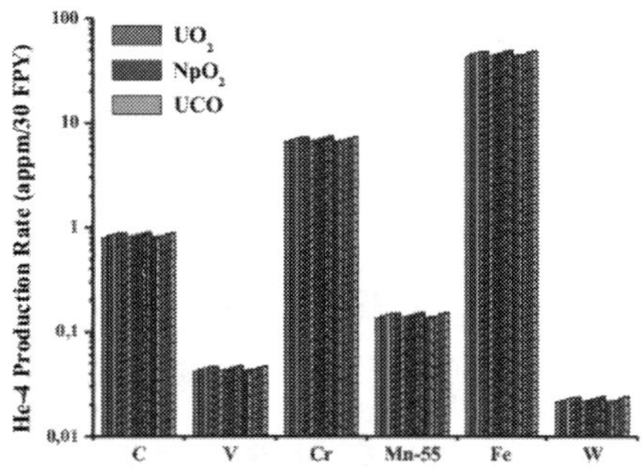

Fig. 5. The contribution of each isotope in the structural material to the total He-4 production rate for the mixture components of UO_2, NpO_2 and UCO (2%, 4%, 6%, 8%, 10%) in the system.

Fig. 5 shows comes from Fe isotope of the biggest contribution and from W isotope of the smallest contribution at the total He-4 production rate in the structural material for used fluids. In Fig. 5, contribution at the total He-4 production rate in the structural material of Fe isotope for used fluids was obtained about 55, $1x10^3$, 7, $3x10^2$, $2x10^3$ times higher than C, V, Cr, Mn-55, W isotopes, respectively. In Fig. 5, the contributions to the total He-4 production rate of Fe isotope in the structural material was found as approximately 85%.

4 Conclusions

It this study was observed that the biggest contribution comes from Cr isotope with approximately 77% value for the total tritium production rate and from Fe isotope with approximately 80-99% between value for the total proton, deuterium, He-3 and He-4 gas production rates in used the structural material. A minimum of gas production is desired, because gas production reduces reactor life and accelerates deterioration of the structural material. In conclusion, the radiation damage parameters obtained according to each isotopes of the structural material in the structural material with 93% $Li_{20}Sn_{80}$ + 5% SFG-PuO_2 + 2% UO_2 fluid showed the best performance of the fifteen tested fluids.

References

1. S. Şahin, M. Übeyli, Energy Convers. Manage., **46**, 3185 (2005)
2. H. M. Şahin, Ann. of Nucl. Energy, **34**, 861 (2007)
3. B. Şarer, et al., Fusion Sci. Technol., **52**, 107 (2007)
4. M. Günay, et al., Ann. of Nucl. Energy, **38(12)**, 2757 (2011)
5. M. Günay, et al., Ann. of Nucl. Energy, **55**, 292 (2013)
6. M. Günay, et al., Ann. of Nucl. Energy, **63**, 157 (2014)
7. IAEA, International Atomic Energy Agency, IAEA-TECDOC-1349 (2003)
8. M. Piera, et al., Energy Convers. Manage., **51**, 1758, (2010)
9. A.Ying, et al., APEX Interim Report, (1999)
10. B. Şarer, et al., Ann. of Nucl. Energy, **36**, 417 (2009)
11. M. Günay, et al., Ann. of Nucl. Energy, **53**, 59 (2013)
12. M. B. Chadwick, et al., Nucl. Data Sheets, **107**, 2931, (2006)
13. D. B. Pelowitz, MCNPX User's Manual, Version 2.7.0, LA-CP-11-00438 (2011)
14. J. J. Duderstadt, G. A. Moses, New York: John Wiley&Sons, (1982)
15. J. A. Blink, et al., Lawrence Livermore National Laboratory, UCRL-53559 (1985)
16. M. Perlado, et al., International Atomic Energy, 272, Vienna, (1995)
17. S. Ünalan, Fusion Eng. Des., **38**, 393, (1998)

Calculation of eigenvalues for neutron transport equation using Henyey-Greenstein phase function in slab geometry

Ahmet Bülbül

Osmaniye Korkut Ata University, Faculty of Arts and Sciences, Department of Physics, Osmaniye, Turkey

Abstract. Eigenvalues are obtained for one-dimensional steady-state neutron transport equation in slab geometry using Henyey-Greenstein (HG) phase function. Firstly, HG phase function is inserted into neutron transport equation then eigenvalues are calculated for different values of collision parameters *c* and *t* parameters. All results are calculated for P_9 and U_9 approximation and these results compared each other.

1 Introduction

In neutron transport problems, all scientist suggested valuable approximations and they solved many problems using various methods in different geometries. In the studies such as eigenvalue spectrum, criticality, flux problem, diffusion length, etc. they have also restrict to problem in the case of isotropic, anisotropic and strongly anisotropic. As well known, eigenvalues depend on the c values and the important coefficients such as diffusion length, diffusion coefficient and buckling also depend on the parameter *c*. In the system, as neutrons move complicated paths and these neutrons repeated nuclear collisions to describe an appropriate scattering function (phase function) is quite important in the solution of algorithm [1]. Scattering function gives the knowledge about behaviour of neutrons which can be absorbed or scattered in the reactor and as well known, as the incident neutron energy increases or the mass of interacting nucleus increases, neutron scattering becomes increasingly anisotropic [2]. In such cases, it is also difficult to predict the distribution of the neutrons in the system accurately. Up to now, different phase functions are used for the solution of the neutron transport equation and convenient results are obtained for each problem.

Henyey-Greenstein phase function (HG) is used in several studies to describe stellar light propagation throughout an atmosphere [3-4] and light scattering in the sea-water [5-6]. HG phase function is also used in bio-medical applications by some researchers [7-8]. Recently, HG phase function applied to neutron transport equation in slab geometry by some researchers and they showed that it is convenient to calculate critical half thicknesses [9]. In one of the latest study, Bülbül et al. calculated critical radius using HG phase function in spherical geometry with P_N approximation [10].

In this study, the eigenvalues have been calculated with U_N approximation using HG phase function in slab geometry. The eigenvalues are calculated for the 9th order of U_N approximation. Obtained numerical results are given in the tables and compared P_N approximation. One can see the effect of the scattering parameter *t* on the solution from the tables and can also decide the mathematical derivations are convenient for the neutron transport equation solutions.

2 Theory

In slab geometry, the steady-state time-independent neutron transport equation without sources is given as

$$\mu\frac{\partial\psi(x,\mu)}{dx}+\sigma_T\psi(x,\mu)=\frac{\sigma_S}{2}\int_{-1}^{1}\psi(x,\mu')\,d\mu', \qquad (1)$$

$$-a\le x\le a,\quad -1\le\mu\le 1.$$

where $\psi(x,\mu)$ is the angular flux or flux density of neutrons at position *x* traveling in direction μ, σ_T and σ_S are total and scattering differential cross section, respectively. It is aimed to solve Equation (1) with HG phase function in this study and to do this, scientist use σ_S in terms of HG phase function and it is given as

$$\sigma_S^{HG}(\mu_0)=\frac{\sigma_S(1-t^2)}{4\pi\left(1-2\mu_0 t+t^2\right)^{3/2}} \qquad (2)$$

where σ_S is any non-negative coefficient, the parameter *t* is in the range of $0\le t\le 1$ and $\mu_0=\mathbf{\Omega}\cdot\mathbf{\Omega}'$ is the cosine of the scattering angel,

$$\mu_0=\mu\mu'+\sqrt{1-\mu^2}\,\sqrt{1-\mu'^2}\,\cos(\varphi-\varphi').$$

The steady state transport equation for one-dimensional case can be written when the HG phase function given in Eq. (2) is inserted on the right hand side of Eq. (1),

ahmetbulbul@gmail.com

DOI: 10.1051/epjconf/201612803001

$$\mu\frac{\partial\psi(x,\mu)}{\partial x}+\sigma_T\,\psi(x,\mu)=$$
$$\int_{-1}^{1}\psi(x,\mu')\mathrm{d}\mu'\int_{0}^{2\pi}\frac{\sigma_s}{4\pi}\frac{(1-t^2)}{\left(1-2\mu_0 t+t^2\right)^{3/2}}\mathrm{d}\varphi' \tag{3}$$

Using

$$\int_{0}^{2\pi}\sigma_S^{HG}(\mu_0)\mathrm{d}\varphi'=\frac{\sigma_S}{2}\sum_{n=0}^{\infty}(2n+1)t^nP_n(\mu)P_n(\mu')$$

the equation can be written

$$\mu\frac{\partial\psi(x,\mu)}{\partial x}+\nu\,\psi(x,\mu)=$$
$$\frac{\nu c}{2}\sum_{n=0}^{\infty}(2n+1)t^nP_n(\mu)\int_{-1}^{1}P_n(\mu')\psi(x,\mu')\mathrm{d}\mu' \tag{4}$$

To simplify the derivation of the equations, here a dimensionless space variable such that $\sigma_T x/\nu\rightarrow x$ is defined and ν is the eigenvalues. In order to solve Eq. (4), it is well known that in the U_N approximation the angular flux is expanded in terms of Chebyshev polynomial as

$$\psi(x,\mu)=\frac{2}{\pi}\sqrt{1-\mu^2}\sum_{n=0}^{N}\Phi_n(x)U_n(\mu),$$
$$-a\le x\le a,\ -1\le\mu\le 1 \tag{5}$$

If the neutron angular flux $\psi(x,\mu)$ given in Eq. (5) is inserted into Eq. (4), and the resulting equation is multiplied by $U_n(\mu)$ and integrated over $\mu\in[-1,1]$ using the orthogonality properties and the recurrence relations of Chebyshev polynomials given below

$$\int_{-1}^{1}U_n(\mu)U_m(\mu)\sqrt{1-\mu^2}\,d\mu=\begin{cases}\pi/2, & n=m\\ 0, & n\neq m\end{cases} \tag{6}$$

$$2\mu U_n(\mu)=U_{n+1}(\mu)+U_{n-1}(\mu),\ -1\le\mu\le 1 \tag{7}$$

One can obtain the U_N moments of the angular flux for $n=0$ and $n=1$ respectively

$$\frac{\mathrm{d}\Phi_1(x)}{\mathrm{d}x}+2\nu\,\Phi_0(x)=2\nu c\Phi_0(x) \tag{8a}$$

$$\frac{\mathrm{d}\Phi_0(x)}{\mathrm{d}x}+\frac{\mathrm{d}\Phi_2(x)}{\mathrm{d}x}+2\nu\,\Phi_1(x)=2\nu ct\Phi_1(x) \tag{8b}$$

In order to obtain the eigenvalue spectrum, a well-known solution is employed of the form [11],

$$\Phi_n(x)=A_n(\nu,t)e^{\sigma_t x/\nu} \tag{9}$$

As well known in U_N approximation, the discrete and continuum ν eigenvalues can be obtained by setting $A_{N+1}(\nu,t)=0$ for various values of c and t. For instance, one can obtain the eigenvalues for U_1 approximation letting $A_2(\nu,t)=0$,

$$\nu_k=\pm\frac{1}{2\sqrt{(1-c)(1-ct)}} \tag{10}$$

From obtained eigenvalues, ν_k, $k=1,\ldots,N+1$, the general solution of the flux moments for odd numbers of N, i.e.

$$\Phi_n(x)=\sum_{k=1}^{\frac{N+1}{2}}\alpha_kA_n(\nu_k,t)\left[\exp\left(\frac{\sigma_Tx}{\nu_k}\right)+(-1)^n\exp\left(-\frac{\sigma_Tx}{\nu_k}\right)\right] \tag{11}$$

where the coefficients α_k can be determined from the physical boundary conditions of the system, and the parity relation of $A_n(-\nu,t)=(-1)^n A_n(\nu,t)$ is used. The other way to compute ν eigenvalues is to use the determinant of the coefficients matrix of $A_n(\nu,t)=0$, that is

$$[\mathbf{M}(\nu)]\mathbf{A}(\nu,t)=0 \tag{12}$$

where $\mathbf{A}(\nu,t)$ is the column vector given by $[A_0(\nu,t), A_1(\nu,t), \ldots, A_{N+1}(\nu,t)]^T$ and $\mathbf{M}(\nu)$ is a $(N+1)\,x\,(N+1)$ square matrix. If one solves the equation det $\mathbf{M}(\nu)=0$ for any order $N+1$, it is clear that same results are obtained with the results of $A_{N+1}(\nu,t)=0$.

Table 1. The eigenvalues for $c=1.01$ and $c=1.2$ and different values of t parameters

t	$c=1.01$		$c=1.2$	
	P_9	U_9	P_9	U_9
0.00	5.750539871	5.750539871	1.198265011	1.198253431
	0.944113891	0.916387696	0.947136093	0.920500197
	0.771095320	0.745729814	0.780103118	0.755077336
	0.503853642	0.478003149	0.515668192	0.488753622
	0.175087652	0.166360374	0.180883259	0.171664714
0.25	6.649411961	6.649411961	1.419574341	1.419569641
	0.948526547	0.922518128	0.954316077	0.930462699
	0.779202263	0.752627310	0.792491581	0.765337133
	0.510727741	0.484687660	0.525352617	0.498290747
	0.177697036	0.168655888	0.184263675	0.174579568
0.50	8.162327211	8.162327211	1.820983261	1.820981761
	1.001414356	0.986441109	1.040978655	1.031656343
	0.825302001	0.790568479	0.852946480	0.815217242
	0.542685181	0.514860964	0.567356918	0.537899614
	0.189058191	0.178697541	0.199060779	0.187562237
0.75	11.61594351	11.61594351	3.092234901	3.092234871
	1.421065877	1.418411784	1.921867006	1.921529396
	1.019227138	0.981682037	1.115927742	1.085095545
	0.680119610	0.640162449	0.755743458	0.708804312
	0.238384902	0.222912740	0.268552301	0.249212801
1.00	97.39065321	95.94929736	4.869532643	4.797464868
	86.50633604	84.12535328	4.325316833	4.206267664
	67.94095710	65.48607339	3.397047842	3.274303670
	43.33953939	41.54150130	2.166976971	2.077075065
	14.88743390	14.23148383	0.744371695	0.711574191

Table 2. The eigenvalues for $c=1.5$ and $c=1.8$ and different values of t parameters

t	$c=1.5$		$c=1.8$	
	P_9	U_9	P_9	U_9
0.00	0.689136251	0.689015261	0.502894771	0.502580591
	0.949871825	0.924354710	0.951455771	0.926634470
	0.789325683	0.764749794	0.795051896	0.770787789
	0.530823974	0.502907528	0.542085776	0.513738785
	0.190431649	0.180422216	0.200228911	0.189458321
0.25	0.842290181	0.842245221	0.630061281	0.629971541
	0.963703864	0.943353644	0.975481641	0.959025451
	0.811700416	0.783146576	0.830349978	0.799692774
	0.547487352	0.519336062	0.568444829	0.539728381
	0.195446003	0.184606249	0.207622727	0.195445564
0.50	1.166373971	1.166373301	0.961683681	0.961781071
	1.168140795	1.165643510	1.604543375	1.604406796
	0.894552174	0.854498094	0.930864256	0.893691006
	0.609460179	0.577613102	0.655157243	0.621058822
	0.217087207	0.203336724	0.238485840	0.221725882
0.75	------	------	------	------
	------	------	------	------
	1.398498024	1.386173187	1.119480408	1.064924851
	0.908153798	0.849845608	4.327527471	4.327564751
	0.335331613	0.305898020	0.446009951	0.395268803
1.00	1.947813057	1.918985947	1.217383157	1.199366218
	1.730126733	1.682507066	1.081329210	1.051566914
	1.358819137	1.309721468	0.849261962	0.818575918
	0.866790788	0.830830026	0.541744242	0.519268766
	0.297748678	0.284629677	0.186092924	0.177893548

3 Results and Discussion

The eigenvalue problem is studied using U_N approximation for the steady-state neutron transport equation without sources. To solve problem HG phase function is inserted into transport equation and eigenvalues are obtained for U_9 and P_9 approximation. All computations are carried out using Maple software and the total macroscopic cross section is assumed to be its normalized value, $\sigma_T = 1\ \mathrm{cm}^{-1}$. The eigenvalues can be calculated by setting $A_{N+1}(\nu,t) = 0$ for various values of c and t. Numerically calculated eigenvalues for $c > 1$ are given in the Table and one pair of the roots is observed purely imaginary the others are real for increasing t parameters from 0.00 to 0.75. The whole eigenvalues for $t=1$ are real and for $c=1.5$, $c=1.8$, $t=0.75$, physically all eigenvalues are existing but mathematically it could not be calculated. We can say that HG phase function works for the calculation of the eigenvalues in slab geometry for neutron transport equation. The aim is in this study that to show applicability of HG phase function to neutron transport equation in slab geometry using U_N approximation and to calculate eigenvalues which can be used for the calculation of criticality, scalar flux, albedo problem, etc.

References

1. W.M. Stacey, *Nuclear reactor physics*. Wiley-Vch Verlag GmbH & Co.KGaA, Weinheim, (2007)
2. H. Brockmann, Proc. 8th Symp. Fusion Techonology, Netherlands, 17-21, (1974)
3. L.G. Henyey, J.L. Greenstein, Astrophys. J. **93**, 70-83, (1941)
4. K. Kamiuto, Journal of Quantitative Spect. and Radiative Trans. **37**, 411-413, (1987)
5. V.I. Haltrin, Applied Optics. **41**, 1022-1028,(2002)
6. G. C. Pomraning, Journal of Quantitative Spect. and Radiative Transfer, **39**, 109-113(1988)
7. L.O. Reynold, N. J. McCormic, Journal of the Optical Society of America, **70**, 1206-1212, (1980)
8. P. Liu, Phys. Med. Biol. **39**, 1025-1036, (1994)
9. H. Öztürk, Kerntechnik, **78**, 447-453,(2013)
10. A. Bülbül, F. Anli, Kerntechnik, **80**, 2, 161-166, (2015)
11. B. Davison, *Neutron transport theory*, London, Oxford University Press, (1958)

Discrete ordinates (S_N) method for the first solution of the transport equation using Chebyshev polynomials

Hakan Öztürk

Osmaniye Korkut Ata University, Department of Physics, Osmaniye, Turkey

Abstract. First estimates for the numerical solution of the one-dimensional neutron transport equation for one-speed neutrons in a finite homogeneous slab is studied. The neutrons are assumed to be scattered isotropically through the medium. Then the discrete ordinates form of the transport equation is solved for the eigenvalue spectrum using the Chebyshev polynomials of second kind in the neutron angular flux. Therefore, the calculated eigenvalues for various values of the c_0, the mean number of secondary neutrons per collision, are given in the tables using the Gauss-Chebyshev quadrature set.

1 Introduction

In the studies about neutron transport through any medium, the scattering of the neutrons should be taken into considerations carefully. Since the neutrons starts the fission in the system, it is important not to lose them to continue the fission chain reaction. Then, when the fission chain reaction begins to increase, it should be taken into a control to fix the power of the reactor. In other words, all reactors are wanted to produce constant energy, i.e. the multiplication factor is desired to be equal to one. This constant power situation can be sensed as isotropic scattering of the neutrons in the system. Therefore, it is still worth to study the isotropic scattering of the neutrons in transport theory.

There are many methods in literature to solve the discrete ordinates (S_N) transport equation numerically. Among them, the Monte Carlo (MCNP) and the source iteration (SI) techniques are extensively used in the solution algorithm of the transport equation. While the MCNP method is one of the first methods, the SI method is reported to converge rapidly for optically thin and slowly for optically thick problems [1,2]. Therefore, alternative methods are always looking for the numerical solution of the S_N transport equation.

In this study, the numerical solution of the S_N transport equation is investigated by using the second kind of Chebyshev quadrature sets. In the solution algorithm of the problem, first an analytic particular solution is derived by applying the separation of variables and then transport equation is converted in a discrete ordinates form which can be solved using Chebyshev quadrature.

DOI: 10.1051/epjconf/201612803002

2 Method

The stationary transport equation with isotropic scattering and constant neutron source can be written for one-speed neutrons [3],

$$\mu \frac{\partial \psi(x,\mu)}{\partial x} + \sigma_T \psi(x, \ ) = \frac{\sigma_{S0}}{2} \int_{-1}^{1} \psi(x,\mu')\mathrm{d}\mu' + \frac{Q_0}{2} \quad (1)$$

where $\psi(x,\mu)$ is the angular flux of the neutrons at position x travelling in direction μ, cosine of the angle between the neutron velocity vector and the positive x-axis. σ_T is the total macroscopic cross-section, σ_{S0} is the differential scattering cross-section corresponding to isotropic scattering and Q_0 is the internal source. In order to solve Eq. (1), the integral term in this equation can be expressed as the integral transform with Chebyshev quadrature;

$$\int_{-1}^{1} \psi(x,\mu')\mathrm{d}\mu' = \int_{-1}^{1} \sqrt{1-\mu'^2}\psi(x,\mu')\frac{\mathrm{d}\mu'}{\sqrt{1-\mu'^2}} \cong \sum_{n=1}^{N} \frac{\psi_n(x)\omega_n}{\sqrt{1-\mu_n^2}} \quad (2)$$

Using Eq. (2) in Eq. (1), one can rearrange the discrete ordinates S_N equations for the numerical solution,

$$\mu_m \frac{\mathrm{d}\psi_m(x)}{\mathrm{d}x} + \sigma_T \psi_m(x) = \frac{\sigma_{S0}}{2} \sum_{n=1}^{N} \frac{\psi_n(x)\omega_n}{\sqrt{1-\mu_n^2}} + \frac{Q_0(x)}{2} \quad (3)$$

where ω_n is the Gauss-Chebyshev quadrature weights or weighting factor for direction μ_n, i.e. the roots of the Nth order Chebyshev polynomials of second kind. Therefore, the roots and the weighting factors of the Chebyshev polynomials of second kind can be given as respectively [4];

$$\mu_m = \cos\left(\frac{m\pi}{N+1}\right) \quad (4)$$

$$\omega_m = \frac{\pi}{N+1}\sin^2\left(\frac{m\pi}{N+1}\right) \quad m = 1,\ldots,N. \quad (5)$$

The general solution of Eq. (3) can be referred to as,

$$\psi_m(x) = \psi_m^p(x) + \psi_m^h(x) \quad (6)$$

where $\psi_m^p(x)$ and $\psi_m^h(x)$ denote the particular and the homogeneous solutions of Eq. (3), respectively. A spatially constant particular solution can easily be got from Eq. (3) with,

$$\alpha_N = \sum_{n=1}^{N} \frac{\omega_n}{\sqrt{1-\mu_n^2}} \quad (7)$$

$$\psi_m^p(x) = \frac{Q_0}{\sigma_T(1-c_0\alpha_N)}, \qquad 0 \le a \le x, \qquad 1 \le m \le N \quad (8)$$

where $c_0 = \sigma_{S0}/\sigma_T$. The homogeneous solution $\psi_m^h(x)$ of Eq. (3) can be determined using the method of separation of variables. Therefore, the homogeneous solution can be written in the form of [5],

$$\psi_m^h(x) = H_m(\nu)\exp(\sigma_T x/\nu), \qquad 0 \le a \le x, \qquad 1 \le m \le N \quad (9)$$

One can obtain an expression for the angular part of the neutron angular flux by using Eq. (9) in Eq. (3),

$$H_m(\nu) = \frac{\nu c_0}{2(\nu+\mu_m)} \sum_{n=1}^{N} \frac{H_n(\nu)\omega_n}{\sqrt{1-\mu_n^2}} \quad (10)$$

and the function $H_m(\nu)$ is normalized by,

$$\sum_{n=1}^{N} \frac{H_n(\nu)\omega_n}{\sqrt{1-\mu_n^2}} = 1 \quad (11)$$

In most of the studies in transport theory, the first step for those problems is to find the eigenvalue spectrum. For this purpose,

Eq. (10) is multiplied by $\left(\omega_m / \sqrt{1-\mu_m^2}\right)$ from both sides and then summed over all m:

$$\sum_{m=1}^{N} \frac{\nu c_0}{2(\nu+\mu_m)\sqrt{1-\mu_m^2}}\omega_m = 1,\ \nu \neq -\mu_m \quad (12)$$

Eq. (12) is referred to as the dispersion relation and the roots $\nu_k, 1 \le k \le N$, of Eq. (12) are the eigenvalues of the S_N equations and they are lying symmetrically about the origin for any c_0 satisfying $0 \le c_0 \le 1$.

3 Numerical results and discussion

The one-speed and one-dimensional transport equation is investigated for the eigenvalue spectrum using S_N method with Gauss-Chebyshev (II. kind) quadrature. For this purpose, first the transport equation in the form of the integro-differential is reduced to S_N form by applying the integral transform with the even-order Gauss-Chebyshev quadrature set. Then, by solving the homogeneous part of the related equation, an analytic expression for the eigenvalues is obtained in Eq. (12). The roots $\nu_k, 1 \le k \le N$ of Eq. (12) are the eigenvalues of the S_N equations, i.e. Eq. (3).

In Table 1, calculated eigenvalues are given for various values of the $c_0 < 1$ and $c_0 > 1$. These eigenvalues are so computed to use different studies related with transport theory.

Table 1. Eigenvalue spectrum

N	$c_0 = 0.30$	$c_0 = 0.60$	$c_0 = 0.99$	$c_0 = 1.20$	$c_0 = 2.00$
2	±0.586037175	±0.740549631	±1.564262966	±1.682828200i^*	±0.554257184i
4	±0.340237412	±0.379634162	±0.442276619	±0.476909013	±0.565264944
	±0.872064483	±1.016288788	±2.732719404	±1.309461630i	±0.457687467i
6	±0.225368072	±0.228117069	±0.231529477	±0.233286929	±0.239447540
	±0.650558027	±0.667286891	±0.679729369	±0.684024964	±0.693422727
	±1.125526328	±2.137618820	±1.245771884i	±0.933040613i	±0.554787056i

$^*i = \sqrt{-1}$

References

1. E.W. Larsen, Nucl. Sci. Eng. **82**, 47-63, (1982)
2. Los Alamos National Laboratory, *Physics Division Progress Report*, Page 70 (1995)
3. K.M., Case, P.F., Zweifel, *Linear Transport Theory* (Addison-Wesley Company, 1967)
4. F., Scheid, *Numerical Analysis* (McGraw Hill, 1988)
5. G.I., Bell, S., Glasstone, *Nuclear reactor theory*, (New York, VNR Company, 1972)

EPJ Web of Conferences **128**, 04001 (2016)
TESNAT 2016
DOI: 10.1051/epjconf/201612804001

Cross section calculations of medical ^{103}Pd radioisotope using α and ^{3}He induced reactions

Bayram DEMIR[1], *Ismail Hakkı* SARPÜN[2], *Yunus Emre* DOGAN[1]

[1]Istanbul University, Department of Physics, Istanbul, Turkey
[2]Afyon Kocatepe University, Department of Physics, Afyonkarahisar, Turkey

Abstract. One of the most popular radioisotopes used in the prostate brachytherapy is Palladium-103 (^{103}Pd). The radioactive plaque is sewn onto the eye as to cover the intraocular tumor shadow with a 2-3 mm margin. These plaques are temporary and radiation is continuously delivered over 5 to 7 days. At the end of treatment, the plaque is removed from eye. In this study, production cross–section calculations of ^{103}Pd radionuclide used in brachytherapy produced by ^{101}Ru(α,2n), ^{100}Ru(α,n), ^{102}Ru(^{3}He,2n) and ^{101}Ru(^{3}He,n) reactions have been investigated in the different incident energy range up to 35 MeV. Two-component Exciton model and Generalized Superfluid model of the TALYS 1.6 code used to perform calculations and calculation results were compared with experimental results reported in the literature.

1 Introduction

One of the most popular radioisotopes used in the prostate brachytherapy is Palladium-103 (^{103}Pd). It is chosen for brachytherapy of because of the low energy photons of 20.5 KeV. Thus, it is possible to minimize doses to normal structures thanks to their low energy with rapid dose fall-off. On the other hand their half-life's (17 days) are very suitable for the brachytherapy. Uveal melanoma can be treated using radioactive ^{103}Pd. Radioactive ^{103}Pd is produced in the form of seed. ^{103}Pd seeds are placed into a non-radioactive metallic plaque to apply the uveal melanoma. The radioactive plaque is sewn onto the eye as to cover the intraocular tumor shadow with a 2-3 mm margin. These plaques are temporary and radiation is continuously delivered over 5-7 days. At the end of treatment, the plaque is removed from eye (1).

Production of medical isotope is an important and constantly evolving issue. Cyclotrons and reactors are used for radionuclide production purposes. The cross–section data for nuclear reactor production are generally well–known and can be satisfactorily reproduced by nuclear model calculations. Theoretical models of nuclear reactions are generally needed to get the prediction of the reaction cross–sections, in a specific manner if the experimental measurements are unobtainable or are improbably to be produced because of the experimental difficulties. In this study, production cross–section calculations of ^{103}Pd radionuclide used in brachytherapy produced by ^{101}Ru(α,2n), ^{100}Ru(α,n), ^{102}Ru(^{3}He,2n) and ^{101}Ru(^{3}He,n) reactions have been investigated in the different incident energy range up to 35 MeV. Two-component Exciton model and Generalized Superfluid model of the TALYS 1.6 code used to perform calculations and calculation results were compared with experimental results reported in the literature.

2 Calculation method used in nuclear reactions

The production cross–sections of ^{103}Pd radionuclide used in prostate and uveal melanoma treatments has been calculated using two–component exciton and generalised superfluid models of the TALYS 1.6 have been used to PEQ reaction calculations. The PEQ reactions were considered by the two–component exciton model. In the two–component model of the TALYS code, the neutron and proton type of the produced particles and holes is explicitly followed all through the reaction. The generalised superfluid model of TALYS 1.6 code takes superconductive pairing correlations into account according to the Barden–Cooper–Schrieffer Theory which is described by a phase transition from a superfluid behaviour at low energy, where pairing correlations strongly influence the level density, to a high energy region which is characterized by the Fermi model. In this model there is no unsolved problem for the parameterization (2).

3 Results and discussion

The production cross–sections of ^{103}Pd radionuclide were calculated as a function of different incident energies up to 35 MeV by using the TALYS 1.6. The calculated and experimental cross–sections of ^{101}Ru(α,2n), ^{100}Ru(α,n), ^{102}Ru(^{3}He,2n) and ^{102}Ru(^{3}He,2n) reactions have been plotted in Figs. 1–4.

DOI: 10.1051/epjconf/201612804001

All experimental results used in this present study were taken from the EXFOR library (3).

The calculated cross–sections of $^{101}Ru(\alpha,2n)$ reaction was compared with the experimental results in Fig. 1. The energy range of α particle was selected between 15 and 23 MeV to make a comparison with the results reported in the literature. Both model calculations are in good agreement with the experimental values except for 15 MeV incident energy.

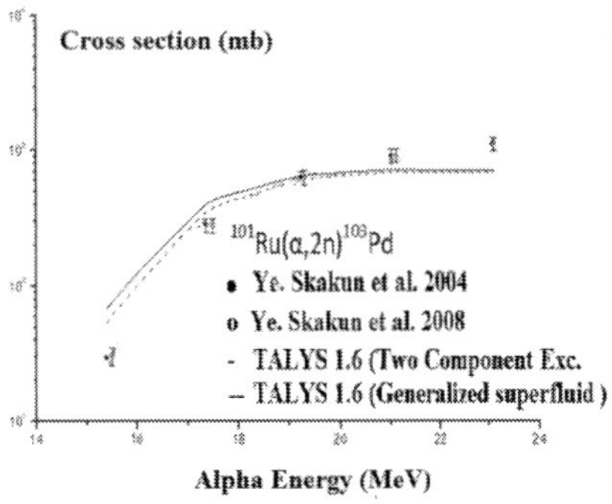

Fig. 1. The comparison of calculated cross–sections of $^{101}Ru(\alpha,2n)$ reaction with the experimental values reported in the literature.

For the 100Ru(α,n) reaction, the comparison of experimental and theoretical results was given in Fig. 2. The energy range of α particle was selected between 11 and 25 MeV to make a comparison with the results reported in the literature. Both model calculations are in good agreement with the experimental data except for 18-25 MeV deuteron incident energy.

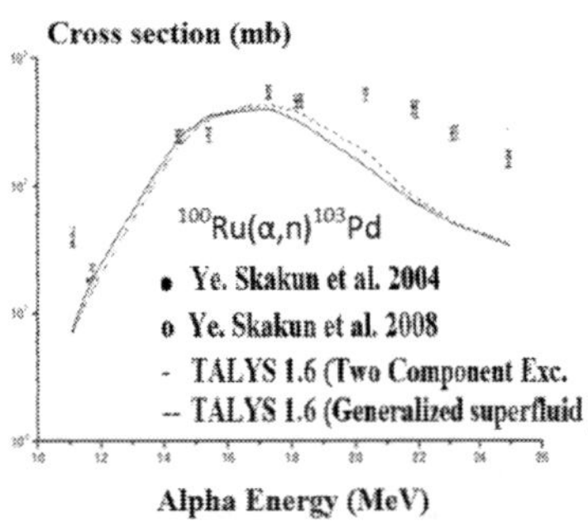

Fig. 2. The comparison of calculated cross–sections of $^{100}Ru(\alpha,n)$ reaction with the experimental values reported in the literature.

The experimental and theoretical results on the production cross–sections of ^{103}Pd radionuclide via $^{102}Ru(^{3}He,2n)$ reaction was given in Fig. 3. The energy range of ^{3}He was selected between 14 and 34 MeV to make a comparison with the results reported in the literature. All model calculations are significantly disagree with the experimental data.

And finally, the experimental and theoretical results on the production cross sections of ^{103}Pd radionuclide via $^{102}Ru(^{3}He,n)$ reaction was given in Fig. 4. The energy range of ^{3}He was selected between 15 and 35 MeV to make a comparison with the results reported in the literature. Both model calculations are in good agreement with the experimental data except for 23-35 MeV ^{3}He incident energy.

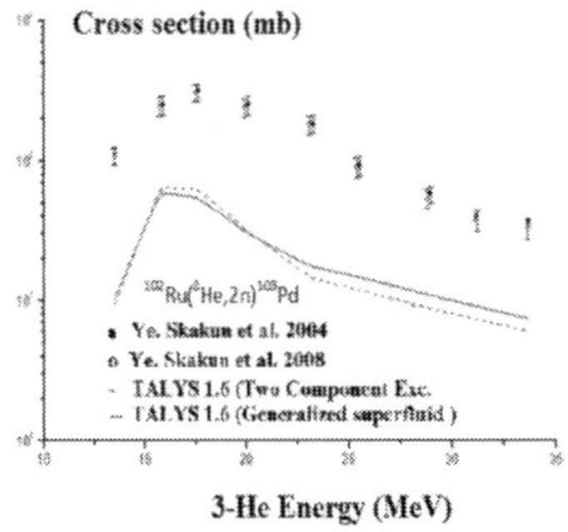

Fig. 3. The comparison of calculated cross sections of $^{102}Ru(^{3}He,2n)$ reaction with the experimental values reported in the literature.

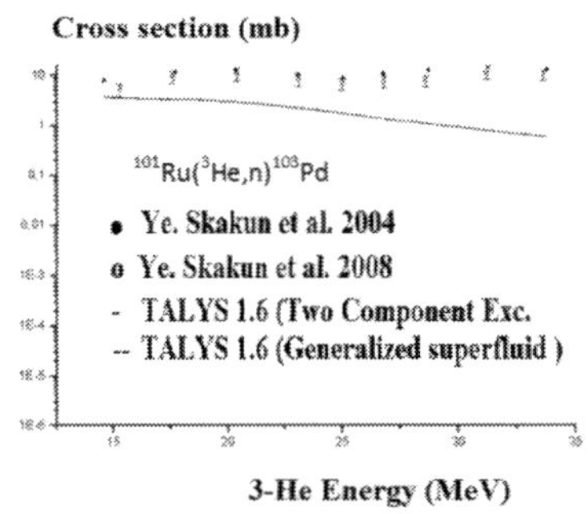

Fig. 4. The comparison of calculated cross–sections of $^{101}Ru(^{3}He,n)$ reaction with the experimental values reported in the literature.

Conclusion

The results can be summarized and concluded as follows:

1) The cross–section results calculated with TALYS 1.6 computer codes for all reactions are mostly in agreement with the experimental data except for $^{102}Ru(^{3}He,2n)$ reaction.

2) The TALYS 1.6 computer codes option for the production cross–section calculations of ^{103}Pd radionuclide can be chosen, if the experimental data are not available or are improbable to be produced due to the experimental difficulty.

References

1. E.R. Simpson, et al., Brachytherapy, Jan-Feb; 13 (1), 1-14. (2014)
2. A. Koning, S. Hilaire, S. Goriely, TALYS–1.6 A Nuclear Reaction Program,64–67. (2013)
3. Brookhaven National Laboratory, National Nuclear Data Center, EXFOR/CSISRS. (http://www.nndc.bnl.gov/exfor/) (2014).

Assessment of patient dose in medical processes by *in-vivo* dose measuring devices: A review

Nina Tuncel

Akdeniz University- Science Faculty -Department of Physics, Antalya-Turkey

Abstract. In-vivo dosimetry (IVD) in medicine especially in radiation therapy is a well-established and recommended procedure for the estimation of the dose delivered to a patient during the radiation treatment. It became even more important with the emerging use of new and more complex radiotherapy techniques such as intensity-modulated or image-guided radiation therapy. While IVD has been used in brachytherapy for decades and the initial motivation for performing was mainly to assess doses to organs at risk by direct measurements, it is now possible to calculate 3D for detection of deviations or errors. In-vivo dosimeters can be divided into real-time and passive detectors that need some finite time following irradiation for their analysis. They require a calibration against a calibrated ionization chamber in a known radiation field. Most of these detectors have a response that is energy and/or dose rate dependent and consequently require adjustments of the response to account for changes in the actual radiation conditions compared to the calibration situation. Correction factors are therefore necessary to take. Today, the most common dosimeters for patients' dose verification through in-vivo measurements are semiconductor diodes, thermo-luminescent dosimeters, optically stimulated luminescence dosimeters, metal-oxide-semiconductor field-effect transistors and plastic scintillator detectors with small outer diameters.

1 Introduction

There are many steps in the chain of processes which determine the absorbed dose delivered to patients in medicine and each of these steps may introduce uncertainties. It is therefore important to check the absorbed dose which has been delivered in practice by putting dosimeters on the patient's skin or in body natural cavities, by placing devices away from patient for detecting the transit photons or emission photons from patients. The ionometric dosimetry by different systems would be applied. This method is called in-vivo dosimetry (IVD). By growing awareness of the need to monitor the dose delivered to patients the implementation of IVD in clinical practice of radiotherapy is widely considered. By the way, in nuclear medicine, diagnostic/interventional procedures at computed tomography, and fluoroscopy/ interventional radiology the in-vivo dosimetry method seem on the rise.

2 Radiotherapy

The IVD in medicine especially in radiation therapy is a well-established and recommended procedure [1-3] for the estimation of the dose delivered to a patient during the radiation treatment. It provides an independent check of the treatment procedure that aims at detection of possible errors in calculation, patient setup and data transfer. In-vivo dosimetry is recognized as part of the quality assurance programme in radiotherapy. It became even more important with the emerging use of new and more complex radiotherapy techniques such as intensity-modulated (IMRT) or image-guided radiation therapy [4]. While IVD has been used in brachytherapy (BT) for decades and the initial motivation for performing was mainly to assess doses to organs at risk (OAR) by direct measurements, because precise evaluation of OAR doses was difficult without 3D dose treatment planning. With the introduction of 3D image-guided BT, it is now possible to calculate 3D tumour and OAR doses but IVD is very relevant in BT as an independent method for detection of deviations or errors [5].

Comprehensive guidelines for the implementation of in-vivo dosimetry with diodes are published by ESTRO [6] and AAPM [7]. Generally, in-vivo dosimeters are available as passive detectors with delayed readout or as real-time detectors with immediate readout. Although real-time techniques have obvious advantages over passive dosimeters, these are supplied immediate judgement possibility regarding to process. Currently 3D image-based dose distributions provide the possibility to determine patient absorbed dose and dose-volume histograms (DVHs) in-vivo at the new innovative techniques in external radiotherapy [8]. The other in-vivo applications for predicting patient absorbed dose after Proton therapy is PET-CT imaging modality [9,10].

3 Nuclear Medicine

The early in-vivo tracer studies by external gamma ray counting used in nuclear medicine and have been progressively shifted from static measurements to dynamic measurements which are performed over a period of time. The variation of the concentration of radioactivity in any site provides the fundamental relationship to be analyzed. The concentration to be measured may occur in a number of different sites, organs, dilution volumes, or combinations of such functional entities. For example, the prior studies on bone were done using ^{47}Ca and ^{85}S radioactive tracers and applied radionuclide scanning to determine and outline the probable site of lesion involved bone [11]. The first applications of external gamma ray counting or in-vivo tracer studies in medicine were the measurement of the uptake of 131-I by thyroid gland at a specific time following the administration of the dose. And additional information could be obtained by measuring the rate of uptake rather by calculating time concentration curves than the final asymptotic value. Then, understanding kinetics of radionuclides is evaluated in the other cases like liver, kidney and bone structures [12]. Nowadays the patient-specific dosimetry for 131-I thyroid cancer therapy using 124-I Positron emission tomography (PET) images which adapted to 3-dimensional–internal dosimetry (3D–ID) software is used for obtaining spatial distributions of absorbed dose, isodose contours, dose–volume histograms (DVHs), and mean absorbed dose estimates [13]. And for the same iodinetherapy cases in-vivo dosimetry performed using patient-specific 3-dimensional methods for SPECT [14]. The fulfilment of such favourable conditions in actual clinical cases can only be confirmed by in-vivo quantization of the absorbed dose achieved as a result of the administration of radioiodine.

4 Radiodiagnostic

The diagnosis with radiation examinations using x-ray tubes to obtain images is based in this field. IVD concept does not used directly but the patient dose is the subject by direct or indirect measurements. The principal quantities for patient dosimetry in general radiography are the incident air kerma, the entrance surface air kerma and the air kerma–area product. For each patient, the incident air kerma can be determined by calculation from recorded exposure parameters and the measured tube output. The entrance surface air kerma provides direct assessment of patient dose and can be measured by TLDs [15,16]. Radiology is an important technique for the diagnosis of patients and in recent years there has been an increase in the use of radiology as part of the treatment of patients. In general, radiation dose levels in radiology are highest for procedures involving fluoroscopy and computed tomography. An international study into patient dose levels in fluoroscopy and digital imaging techniques has been initiated using dose area product meters. A computerized method of automatically monitoring tube and generator parameters to perform on-line quality assurance and patient dosimetry has been developed [17,18]. Today, with the introduction of digital radiology, it is possible to have automatic systems to collect and archive patient dose data individually, for all the imaging and interventional procedures, in addition to demographic, geometric, and other procedural parameters, as part of the DICOM header or through other DICOM services [19].

5 IVD: Passive and Active Methods

In-vivo dosimeters can be divided into real-time and passive detectors that need some finite time following irradiation for their analysis. Both types of dosimeter require a calibration generally obtained by comparing their response against a calibrated ionization chamber (IC) in a known radiation field. Most of these detectors have a response that is energy and/or dose rate dependent and consequently require adjustments of the response to account for changes in the actual radiation conditions compared to the calibration situation. Correction factors are therefore necessary to take, for instance, changes in field size, source-detector distance, temperature, pressure, and orientation, including the presence of a build-up cap, into account. The presence of a build-up cap is important for detectors used for entrance or exit IVD [4].

Today, the most common dosimeters for patients' dose verification through in-vivo measurements are semiconductor diodes, thermoluminescent dosimeters (TLDs) and optically stimulated luminescence dosimeters (OSLDs), or electronic portal imaging devices. There is also increased application of metal-oxide-semiconductor field-effect transistors (MOSFET) for clinical radiotherapy dosimetry. Diode-based in-vivo dosimetry has been widely explored and reported in the literature [20-24]. Similar investigations have been reported with TLDs [25,26] and with MOSFET dosimeters [27-29]. On the other hand OSLDs potential in medical dosimetry [30] was recognized due to their small size suitable for point measurements, high sensitivity, non-destructive readout and reanalysis and the use of relatively simple readers with a high degree of automation. Technical advances in the production of OSL materials and availability of the reading devices make them a very promising dosimetric tool for medical purposes. Current developments and specific applications of OSLDs in diagnostic radiology and radiation therapy are reconsidered [31-33].

TLDs and OSLDs were the most common choice of passive dosimeters for medical purposes. However, OSLDs have an advantage compared with TLDs [34], since no heating is required for the dose measurement, which makes the reading equipment cheaper, and easier to handle and maintain. These are still widely used in routine work for several reasons. Usually there are a larger number of dosimeters provided for in-vivo measurements which allows for simultaneous dose determination at several positions on the patient. No cable connection between dosimeters and the reader allows their usage on different treatment machines in the department. Moreover, it is not necessary for the hospital to be equipped with the reading unit; instead, readout can be arranged in an outside institution.

The innovative solution of so-called in vivo "transit" dosimetry currently is being implemented in many radiotherapy centers. Usually used to check patient positioning, portal imagers (Electronic portal imaging device; EPID) allow two-dimensional acquisition of the signal transmitted through the patient throughout the entire duration of irradiation. It is then possible to arrive at an estimate of the dose in the patient. The dose thus reconstructed can then be compared to the planned dose to validate the quality of the treatment delivered. This method has a certain number of advantages compared to the "standard" method so measurement of the dose at a point by placing a detector on the skin and opens up a number of prospects with regard to strategies for validation of the dose delivered to the patient. Currently 3D EPID-based dose distributions provide the possibility to determine patient absorbed dose and dose-volume histograms (DVHs) in-vivo at the new innovative techniques in radiotherapy [8]. The other applications for predicting patient absorbed dose after proton therapy is PET-CT imaging modality [10].The later methods that based on images and applied soft ware could be consider as real time in-vivo dosimetry.

6 Conclusions

IVD programs in radiotherapy contributed to increased expertise in radiation dosimetry in clinical environment leading to increased precision on treatment delivery, better detection of systematic error and the prevention accidents in radiotherapy. The innovative approach of so-called in-vivo "transit" dosimetry allows in vivo dosimetry to be extended to other fields of medicine. Studies are also underway on this approach.

References

1. AAPM Report No.46, Med. Phys.Med. Phys. **21**, 581 (1994).
2. M. Essers, B.J. Mijnheer, Int. J. Radiat. Oncol. Biol. Phys. **43**, 245 (1999).
3. IAEA, *Setting Up a Radiotherapy Programme*, , Vienna (2008).
4. B. Mijnheer, et al., Med. Phys. **40**, 1 (2013).
5. K. Tanderup, et al., Med. Phys. **40**, 1 (2013).
6. D. Huyskens, et al., (ESTRO Booklet No.5, Brussels, 2001).
7. AAPM Report No.87, AAPM (2005).
8. B. Mijnheer, et al., Journal of Physics: Conference Series 444, (2013).
9. K. Parodi, **68**, 920–934 (2007).
10. X. Zhu, G. El Fakhri, **3**, 731-740 (2013).
11. D. A. Weber, et al., J Nucl Med., **10**, 8-17 (1969).
12. M. G. Stabin, *Fundamental of Nuclear Medicine Dosimetry*, (Springer, 2008).
13. G. Sgouros, et al., J. Nucl. Med., 45 (8) 1366 (2004).
14. Y. K. Dewaraja, et al., J. Nucl. Med., 46:840–849 (2005).
15. ICRU Report 74, (ICRU, Bethesda, MD, 2006).
16. IAEA, *Technical Reports Series No.* 457, (IAEA, Vienna, Austria, 2007).
17. K. Faulkner, et al., **12**, 228-232 (1990).
18. A. Meghzifene, et al., European Journal of Radiology, **76**, 11–14 (2010).
19. E. Vano, et al., AJR, **200**, 783-790 (2013).
20. C. J. Millwater, A. S. MacLeod, D. I. Thwaites, Br. J. Radiol. **71**, 661–8 (1998).
21. R. Alecu, T. Loomis, J. Alecu, T. Ochran, Med. Dosim. **24**, 5–12 (1999).
22. N. Jornet, M. Ribas, T. Eudaldo, Med. Phys., **27**, 1287–93 (2000).
23. C. Fiorino et. al, Radiother.Oncol., **56**, 85–95 (2000).
24. P. A. Jursinic, Med. Phys., **28**, 1718–26 (2001).
25. T. Loncol, et al., Radiother. Oncol., **41**, 179 (1996).
26. A. Swinnen, J. Verstraete, D.P. Huyskens, Radiother. Oncol., **73**, 89 (2004).
27. R. Ramaseshan, et al., M. Islam, Phys. Med. Biol., **49**, 4031 (2004).
28. C.W. Scarantino, et al., Int. J. Radiat. Oncol. Biol. Phys., **62**, 606–13(2005).
29. A. Cherpak, et al., Med.Phys., **36**, 1672–9 (2009).
30. M.S. Akselrod, et al., Radiat. Meas., **41**, 78 (2006).
31. E.G. Yukihara, S.W.S. McKeever, Phys.Med. Biol., **53**, 351 (2008).
32. E. Yukihara, et al., Med. Phys., **35**, 260 (2008).
33. E.G. Yukihara, et al., Radiat. Meas., **45**, 658 (2010).
34. S.W.S. McKeever, M. Moscovitch, Radiat. Prot. Dosim., **104**, 263–70 (2003).

 DOI: 10.1051/epjconf/201612804003

Monte Carlo based dosimetry for neutron capture therapy of brain tumors

Lilia Zaidi[1], *Mohamed* Belgaid[1], and *Rachid Khelifi*[2]

[1]Faculté de physique USTHB, Laboratoire SNIRM, BP 32 EL-Alia,BEZ Algiers, Algeria.
[2]LPTHIRM, Département de Physique, Université Saad Dahlab, BP 270 Route de Soumaa, Algeria.

Abstract. Boron Neutron Capture Therapy (BNCT) is a biologically targeted, radiation therapy for cancer which combines neutron irradiation with a tumor targeting agent labeled with a boron10 having a high thermal neutron capture cross section. The tumor area is subjected to the neutron irradiation. After a thermal neutron capture, the excited ^{11}B nucleus fissions into an alpha particle and lithium recoil nucleus. The high Linear Energy Transfer (LET) emitted particles deposit their energy in a range of about 10μm, which is of the same order of cell diameter [1], at the same time other reactions due to neutron activation with body component are produced. In-phantom measurement of physical dose distribution is very important for BNCT planning validation. Determination of total absorbed dose requires complex calculations which were carried out using the Monte Carlo MCNP code [2].

1 Introduction

During BNCT, the absorbed dose delivered to the normal tissue and tumor, result from a various field irradiation exposure that depends on the spatial, spectral and angular characteristics of the neutron beam as well as on the geometry and composition of the target. The principle absorbed dose components includes: the alpha particles and lithium ions resulting from the boron neutron capture reaction $^{10}B(\alpha,n)^{7}Li$; recoil protons from fast neutron interactions largely due to the $^{1}H(n,n)^{1}H$, and $^{14}N(n,p)^{14}C$ reactions. All of (n,γ) radiative capture reactions particularly the neutron hydrogen capture reaction $^{1}H(n,\gamma)^{2}H$ were considered [3].

Depending on the position of the tumor, two different neutron beams are commonly used in BNCT: the thermal neutron beam, which limits the treatment to shallow tumors, such as skin melanoma, and the harder epithermal neutron beam as it can penetrate deeper into tissues due to its high energy, it has been used for treating the deep-seated tumors. Since we are interested into deep brain tumors such Glioblastoma multiform GBM, the later beam is the adequate one to be used.

In this study, the Radio biological effectiveness (RBE) dose, the depth dose profiles along the central axis of the Snyder's head phantom and in-phantom parameters have been computed for the proposed neutron beam irradiation.

2 Materials & Methods

2.1 Beam irradiation

The neutron beam from an accelerator based on the $^{7}Li(p,n)^{7}Be$ reaction [4], with 2.3 MeV proton energy and 10 mA current, was considered. The primary neutrons from the $^{7}Li(p,n)^{7}Be$ reaction for proton energy of 2.3 MeV have an average energy of 0.233 MeV which is considered inadequate for treatment with BNCT.

In order to provide an appropriate neutron beam, mainly epithermal neutrons of energy range (0.5 eV < E < 20 keV), a special Beam Shaping Assembly (BSA) must be designed based on the neutron source specifications. A typical BSA includes moderator, reflector, collimator, thermal neutron filter, and gamma filter.

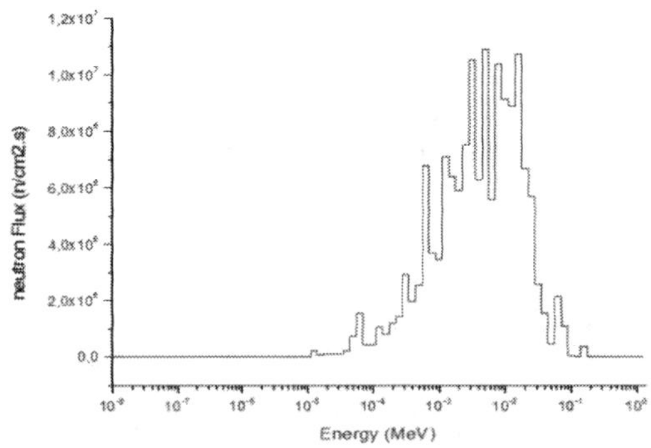

Fig. 1. Neutron beam spectra obtained with the proposed BSA configuration.

In our previous work, detailed studies on materials consisting of the (BSA) was done, where the optimal configuration was chosen as the one can treat tumors in the biggest range of depth and shorter treatment times with best therapeutic ratio. An optimal therapeutic neutron beam presented in *Fig.1*, centered around 10 keV energy, was obtained. In this study, the beam generated from the BSA is considered as source of irradiation.

DOI: 10.1051/epjconf/201612804003

2.2. Dosimetry

An ellipsoidal head phantom based on the modified Snyder shown in *Fig.2*, was considered as a representation of a patient head, the elemental compositions for material of the analytical phantom: scalp, skull and brain have been taken from ICRU 46 [5] of adult head person. The head was positioned at the exit side of the BSA *Fig.3*.

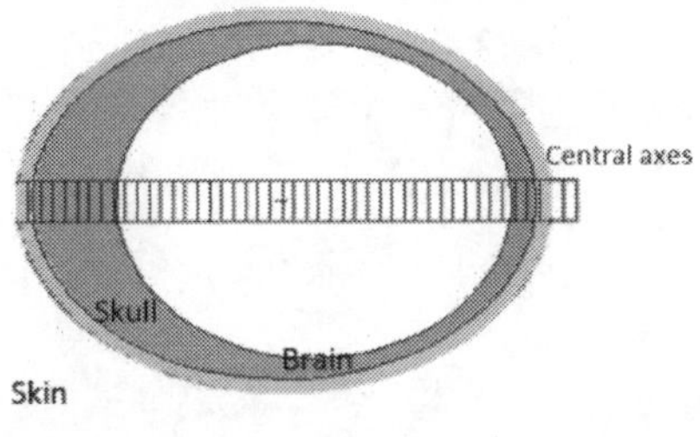

Fig. 2. Snyder head phantom

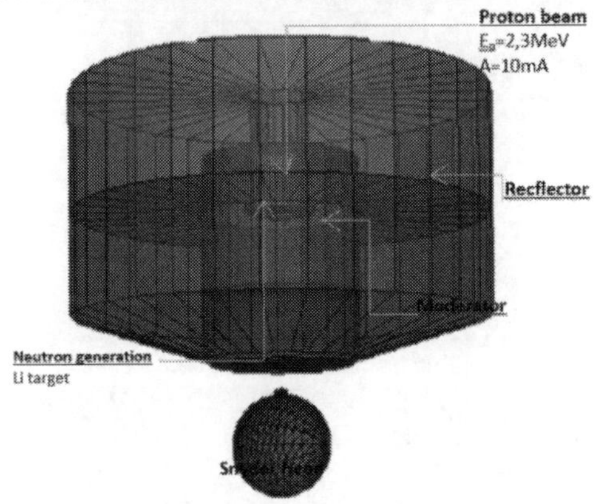

Fig. 3. Schematic view of the BSA and the head phantom configuration simulated for dose calculations.

The absorbed dose from Heavy Charged Particles (HCP) was computed from the neutron flux generated by MCNP for each cell tally in Head phantom. Conversion of particle flux to dose was carried out using the conversion factors for H, ^{16}O, ^{14}N, ^{12}C, ^{31}P and ^{40}Ca in ICRU 63, based on the ENDF/B-VI nuclear data library [6]. Kerma factors for the elements Cl and S were calculated from JENDL-3.2 cross section data and Q-values [7].

The gamma ray flux resulting from both the neutron capture gamma rays and the gamma rays from the Li-target and the BSA were converted into absorbed doses using the dose conversion factors based on mass energy absorption coefficient, described by Seltzer [8].

The weighted total dose is defined as the sum of physical dose components multiplied by the weighting-factors of each dose component in a tissue:

$$D_w = w_g\,D_g + w_B\,D_B + w_N\,D_N + w_{fn}\,D_{fn} \qquad (1)$$

where D_w is the weighted total dose, D_g is the gamma dose, D_B is the absorbed dose due to the boron, D_N is the nitrogen dose and D_{fn} is the fast neutron dose. The weighting factors w_N and w_{fn} were taken as 3.2, w_g was considered to be 1.0 and w_B was considered 1.35 and 3.8 in the normal tissue and tumor, respectively [9].

The ^{10}B concentrations were assumed to be 52.5 ppm for tumor and 15 ppm for healthy tissue. These four dose components have been computed separately along the center line of the beam through the brain using flux to dose conversion factors using the MCNP F4 tally and DE4/DF4 cards.

3 Results and discussions

In-phantom parameters have been calculated in the computational Snyder head phantom model for the particular case of the optimized neutron beam shown in Fig.1. These parameters are: Advantage Depth (AD) (the depth of effective beam penetration at which, the ratio of the dose in the tumour to the dose in the normal tissue, is the maximum); Treatable depth (TD), the depth at which the tumour dose falls below twice the maximum dose to normal tissue, and the treatment time (TT).

Considering that the maximum allowable dose to healthy tissue is 12.5 Gy, TT can be estimated. On the other hand, In BNCT, treatment time can extend up to one hour. Results in Table 1 shows the estimated parameters for the neutron beam generated by the optimized BSA, and the treatment capabilities.

Parameters	Values
Total dose (RBE.Gy/min)	0,519
Max Tumor dose (RBE. Gy/min)	4,31E-01
Mean Brain dose (RBE.Gy/min)	4,140486E-2
Skin dose (RBE.Gy/min)	3,11E-02
Dose ratio or therapeutic gain	4,92
Treatment time TT (h)	1
Treatable depth TD (cm)	7,1
Maximum healthy brain punctual dose 11 RBEGy	4,8

The Fig. 4. shows the depth profile of the dose components rate for healthy tissue and tumour.

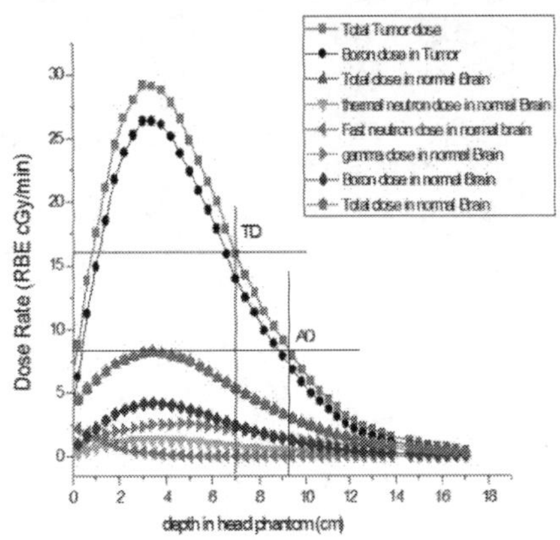

Fig. 4. Distribution of various dose components in the head

In the brain the boron component due to nonspecific boron deposition in healthy tissue dominates. The gammas dose is the second component contribution, mainly due to gammas produced inside the phantom. These two components are the limiting factors in healthy brain while at the tumour tissue the dominant dose component is due to Boron dose. The dose contribution from hydrogen recoil due to fast neutrons reactions on hydrogen and nitrogen still small.

4 Conclusions

In this study, by means of numerical simulations, the RBE dose, depth dose profiles have been computed and the beam quality was also characterized with the help of the dose estimation in the Snyder's head phantom.

The obtained results in this work, for the considered neutron beam source, show good treatment possibilities with short irradiation times. The beam penetration is good: maximum tumor dose at 3.4cm depth; and an advantage depth of 9.2 cm. with the maximized dose ratio 4.92 of tumor to normal tissue. Then the designed beam encompass suitable quality and intensity, it can deliver more than 25.8 RBE Gy to a tumor in one hour for only 10mA proton current.

References

1. IAEA, International Atomic Energy Agency, *IAEA-TECDOC-1223*, (2001).
2. J.F. Briesmeister, RSICC, Oak Ridge National Laboratory, LANL manual, LA-12625-M, Los Alamos, (1997).
3. J. T. Goorley, W. S. Kiger, and R. G. Zamenhof, *Med. Phys.*, **29**, 145–156, (2002).
4. C. L. Lee and X. Zhou, **152**, 1–11, (1999).
5. ICRU 46, International Commission on Radiation Units and Measurements Tech. rep. (1992).
6. Brookhaven National Laboratory, "ENDF/B-VI Summary Documentation," BNL-NCS-17541, (1991).
7. T. Nakagawa, K. Shibata, S. Chiba, T. Fukahori, Y. Nakajima, Y. Kikuchi, T. Kawano, Y. Kanda, T. Ohsawa, H. Matsunobu, M. Kawai, A. Zukeran, T. Watanabe, S. Igarasi, K. Kosako, and T. Asami, "Japanese Evaluated Nuclear Data Library Version 3 Revision 2: JENDL-3.2," J. Nucl. Sci. Technol., **32**, 1259 (1995).
8. J.H. Hubbell, S.M. Seltzer. NISTIR 5632, National Institute of Standards and Technology, (1995).
9. D.T. Ingersoll, et al., in Advances in Neutron Capture Therapy, Vol. I, edited by B. Larsson, J. Crawford, and R. Weinreich, Elsevier, Amsterdam (1997), pp. 95-99

EPJ Web of Conferences **128**, 05001 (2016) DOI: 10.1051/epjconf/201612805001
TESNAT 2016

An investigation on nuclear energy policy in Turkey and public perception

Mehmet Burhanettin Coskun[1], *Banu* Tanriover[2]

[1]Department of Politics and Public Administration, Osmaniye Korkut Ata University, Osmaniye, Turkey
[2]Department of Economics, Osmaniye Korkut Ata University, Osmaniye, Turkey

Abstract. Turkey, which meets nearly 70 per cent of its energy demands with import, is facing the problems of energy security and current account deficit as a result of its dependence on foreign sources in terms of energy input. It is also known that Turkey is having environmental problems due to the increases in CO_2 emission. Considering these problems in Turkish economy, where energy input is commonly used, it is necessary to use energy sources efficiently and provide alternative energy sources. Due to the dependency of renewable sources on meteorological conditions (the absence of enough sun, wind, and water sources), the energy generation could not be provided efficiently and permanently from these sources. At this point, nuclear energy as an alternative energy source maintains its importance as a sustainable energy source that providing energy in 7 days and 24 hours. The main purpose of this study is to evaluate the nuclear energy subject within the context of negative public perceptions emerged after Chernobyl (1986) and Fukushima (2011) disasters and to investigate in the economic framework.

1 Introduction

Quest created by the growing need for sustainable energy in Turkey has started to highlight the various renewable energy needs. This creates the need for nuclear energy of the first actors to trigger the global and technological development as well as establishing and operating direction of the well has now started to be observed to make a functional form that can spread the form of energy production. Nuclear power plants and power centers of the energy system could be used in our country to build off of existing nuclear power plants is clear that the most rational approach is no longer a hidden fact. Brought to life within the context of this reality constitutes the main subject of the Energy Policy and Public Perception survey creation in Turkey.

As an alternative solution that can contribute to the current electricity generation of Turkey's energy needs, the basic aim of the research on nuclear energy in which there is a positive example on the world economic paradigm, social and political returns to be created in the public perception of environmentally compatible systems in nuclear power consumption reveal positive and negative and made public which is to contribute to the reduction of false cause of uncertain thoughts. With this conceptual idea of the benefits of nuclear power use is not only about the national doubts but also with their expectations and perceptions. Globalization in the area where the production of energy spending and economic competition with countries has been diligently calculated deficits and dependencies. It is able to show growth without getting stuck on the negative perception of Turkey will provide the benefits of nuclear energy policy? In this research, descriptive and explanatory methods were used to put forward the findings identified in this line of thinking.

This paper is organized as follows. Section 2 gives an overview about the role and importance of nuclear power plants in the world. Section 3 presents the energy problematical and nuclear energy policies in Turkey. The public perception of nuclear energy in Turkey has been evaluated in the section 4. Conclusions and discussions have been given in the section 5.

2 The role and importance of nuclear power in the World

After the first use of nuclear energy for electricity production in the US, it began to be used in electricity production by England in 1953, by Russia in 1954, by France 1956 and by Germany in 1961. 10 countries in the 1960's and subsequently 10 countries more in 1970's began producing electricity based on nuclear technology. The oil crisis of the early 1970's has increased the demand for nuclear power plants and launched a wave of constructing these plants. Next decade slowdown in the global economy and the decline in fossil fuel prices has restricted the growth of nuclear energy demand. Contrary to this, serious concerns occurred about the safety of nuclear facilities on a number of public by the impact of nuclear power plant accidents. However, some countries have continued construction of the reactor, and this has resulted in a limited increase in nuclear energy [1].

When historical development of nuclear energy is examined; it can be seen that the first use in order to generate electricity in the past in a commercial nuclear power has been in 1954. Nuclear energy was considered to be the largest source of energy on the world as "cheap, clean and safe" after the transition term of oil. Although it is understood that the end of 20th century it is not so easy accomplish it but, it is also an irresistible truth that many developed or developing countries is spreading increasingly to nuclear power plants. One of these countries is Turkey which is planning to get electrical energy from nuclear plants [2].

Currently 17% of the world's electricity needs are obtained by nuclear energy, nuclear technology industry and by the use of production of isotopes it is also an important issue for many other sectors such as in medicine except energy [3]. Nuclear power plants began to be built after the discovery of the energy obtained from the nucleus of nuclear weapons [4]. Nuclear power plants improved in this process due to Britain's, America's and Russia's studies by using it for military purposes in the early 1944's [5]. The first nuclear power plant to produce electricity in the world was founded in Russia in 1954. After Russia, economic giants such as the US, Britain and France built nuclear power plants [6].

Today, nuclear energy is getting widespread use in some parts of the world as a kind of energy. Almost 17% of the electricity need is supplied by nuclear energy with approximately 440 nuclear power plant worldwide. In some countries and regions, a large part of the energy requirements is supplied by the nuclear power [7].

Up to 87% of nuclear power generation is held by OECD countries worldwide [8]. The number of countries which use nuclear power is very limited. There are 31 countries (16%) which have nuclear power in United Nations. The ¾ of the world's nuclear power is generated by US, France, Japan, Germany and South Korea. In addition to this, these countries have nuclear weapons [9] For example this ratio is 78% in France and 20% in US. The alternative energies weren't used such as wind and geothermal during rapid spread of nuclear energy. However, due to the increase in security measures today, nuclear waste and licensing costs, nuclear energy costs increased fairly [10].

On the other hand, Sweden which has 12 and Germany which has 21 nuclear power plants have stopped the new nuclear power plant projects. Both two countries will also close their nuclear power plants until 2020. Austria, Denmark, Italy and Spain, have no nuclear power plant project in their agenda [11] Indeed, Germany also began to criticize policies on nuclear countries such as Belgium to stop the activities of nuclear power plants and it became a very serious discussion.

Germany has requested the closure of two nuclear power plants of Belgium on the grounds of the threats. Electricity generation supplied from nuclear energy plants is ranging from 1.5 % to 2.5% worldwide. In general, the world has witnessed a decline in the production capacity of nuclear power due to the responses from the public, the problems encountered in the storage of waste, concerns and proliferation of nuclear weapons and difficulties encountered in the nuclear energy industry [9].

3 Energy problematical and nuclear energy policies in Turkey

Energy, which is one of the main inputs of industrialisation process, is an important factor of production in order to ensure sustainable development. Approximately 75% of energy demand in the world is provided from fossil fuels. If new energy sources are not derived, it is estimated that fossil fuels will run out after about 70 years and the world's energy crisis will become. Fossil fuels are limited amount and also cause environmental problems due to the increases in CO_2 emission. In the recent years, the strategy independency in energy, which is one of the objectives of national economy recently, dramatic increases in energy demands as a result of industrialisation and technological developments and the fact that energy sources are decreasing rapidly are to turn the countries to the quest for new energy. Turkey, which meets nearly 70 per cent of its energy demands with import, is facing the problems of energy security and current account deficit as a result of its dependence on foreign sources in terms of energy input. It is also known that Turkey is having environmental problems due to the increases in CO_2 emission. Considering these problems in Turkish economy, where energy input is commonly used, it is necessary to use energy sources efficiently and provide alternative energy sources.

These alternative energy sources include in nuclear energy and renewable energy such as wind, wave, biomass, geothermal, hydraulic and hydrogen energy. Due to the dependency of renewable sources on meteorological conditions (the absence of enough sun, wind, and water sources), the energy generation could not be provided efficiently and

 DOI: 10.1051/epjconf/201612805001

permanently from these sources. At this point, nuclear energy as an alternative energy source maintains its importance as a sustainable energy source that providing energy in 7 days and 24 hours. Therefore, just in being the 21st century, new energy generation projects that considered the global climate changes and sustainable development, have become important. Especially, oil crisis occurred in the beginning of 1970's accelerated these initiatives and provided to come to the forefront of nuclear energy as a reliable source. Consequently until the second half of the 1980's many nuclear reactors were installed and put into operation. However by the end of 1980's demand for nuclear energy decreased and during 1990's this demand became stable. Three Mile Island (1979, USA) and Chernobyl (1986, former USSR) nuclear accidents, the recession in the world economy, and penetration of natural gas to the energy generation market are shown among results [12]

It is also clear that almost all of the natural gas and liquid fuels and 30% of coal used for electricity generation in Turkey are imported from other countries. On the other hand, in case all of Turkey's renewable energy sources like hydroelectric, wind, solar, geothermal and biomass are used for generating electricity, the half of the Turkey's electricity consumption demand of 500 billion kWh could be provided from these sources alone. Therefore, the important of nuclear energy as an alternative energy source is increasing day by day to be reducing of dependence on foreign sources in energy, to be reducing of emissions in the result from using fossil fuels, to be increase the international competitive power, to offer alternative sources to transitory supply resulting from the use of renewable energy sources. The other reasons why nuclear energy is preferred world-wide and in Turkey for electricity generation are following [13]:

-Nuclear power plants have the continuous availability in generating electricity. Otherwise, the generated electricity depends on the right meteorological conditions in renewable energy sources power plants and the quality and composition of coal in coal fired plants and the quantity of reserves in fuel and natural gas power plants. Nuclear power plants do not require specific locations in order to generate large amounts of power.

-While considering nuclear fuel cycle, nuclear power plants are the eco-friendly option for greenhouse gas emissions to the environment. There are no released gases while nuclear power plant operated just as in released by combustion of fossil fuels like carbon monoxide, carbon dioxide, sulphur dioxide and nitrogen oxides gases. Nuclear energy stations do not emit criteria pollutants or greenhouse gases when they generate electricity. Because of this, nuclear energy has an inevitable role in reducing the greenhouse gas emissions caused to the climate change. Nowadays, the emitted greenhouse gases amount is reduced approximately by 17 % annually due to the electricity generated in nuclear power plants. If the fossil fuel power plants are used instead of nuclear power plants, the emitted carbon gas amount would be 1,2 billion tons.

Table 1. Changes in costs of unit electrical production when fuel prices and ratios doubled in operations

Fuel Type	Used Fuel Ratio	Changes in Costs When Fuel Prices Doubled
Coal	77%	31% increase
Natural Gas	90%	66% increase
Nuclear Energy	30%	9% increase

Source: Energy Resources International (ERI), Washington, May 2011

-As shown in Table 1, the share of nuclear fuel in operating cost is around 30 % (while this is 77 % for coal power plants and 90 % for natural gas power plants). Directly, the fluctuations in nuclear fuel prices is less impressed the cost of electricity production in nuclear power plants. When the fuel prices doubled, the effect of that cost increase the total operating cost in 66 % in natural gas and in 31 % in coal.

-Nuclear power plants contribute to country's economic development by providing new employment areas.

-The generated energy from nuclear power plants diversifies the energy resources portfolio of country.

4 Public perception of nuclear energy in Turkey

Increasing world population, natural resource consumption trends and developments and the need for industrial machinery is directly pushing forward the use of limited fossil fuels. It can be seen that renewable energy sources move in a linear relationship with the establishment of nuclear power plants. Now the developed world countries are making an effort to get ahead of coming to dangerous levels of greenhouse gas emissions. All member countries of the UN were encouraged to be won a strong action plan with a consensus and accordingly the use of technology to prevent the increase of fossil fuel use within a program in order to apply needs increasing gradually.

In December 2015, members of the United Nations countries had participation and prepared a preliminary protocol by consensus in Paris. After that on 22nd April 2016 unanimous vote of the 195 representatives of UN countries accepted "climate change" with the signatures as a protocol in New York. The protocol seeks to reduce this global warming almost 2 degrees for next century and 100 billion US dollars budget allocation was decided for developing countries. Thus, emission attack against the car company that we see recently and against greenhouse gas emissions to be taken by the countries which took the lead may be in tightening measures and global warming may also be prevented.

At this point, while the global warming and greenhouse gas emissions were discussed, the issue remained almost alone on our fossil fuel dependence on energy imports and our review has required more serious public perception and people witness to the government policy to support efforts to strengthen the energy perception on this issue. We can observe that the nuclear issue is being mentioned by both by scientists and politicians in addition to appearance of the benefits and damages raised from the use of nuclear energy and to eliminate the negative perception to Turkey which was affected by the Chernobyl nuclear accident in 1986.

Despite the expansion of the technological and economical use of maturation, the perception of nuclear energy on people stands in a different place among other energy sources. Many factors has contributed to this idea such as military purposes, the possibility that can be used to make weapons, technical complexity, long-term effects of nuclear waste, complex insurance requirements and safety law, the results associated with potential accidents, many health effects and high initial investment costs of exposure to ionizing radiation. Nuclear energy can be better understood by the help of well assimilation of these issues [1].

It is scientifically proven today that non-renewable energy sources such as oil, natural gas and coal is projected to be used respectively in 54-year period, 64 years and 112 years and decreases day by day. While the amount of total energy production from fossil fuels in the world is 10,905.9 Mtoe, the consumption amount stood at 10,689 Mtoe [14]. This is where the nations' economic, social and in accordance with the nature of nuclear energy will occur as general acceptance of a result on a basic social perception. It is important that this perception to be presented in the form of campaigns with rational and analytical grounds.

5 Conclusion

Global warming, rising energy costs and economic external dependence have been very important topics day by day in the world. These discussions in many of the developing countries and supranational organizations continue to create different reflexes. The main aim of these reflexes is to make a cleaner and more liveable world. Now it became mandatory to get a rational solution on growing energy needs and rising industrialization levels. Turkey is a developing country which has a very large amount of import. Current energy problems can be solved by a perception which can be creates from the thoughts of seasonal efficiency of renewable energy and minimizing the security issues on infrastructure of nuclear power plants. So it will be easy to build up the efficient use of nuclear power plants planned in Turkey. By the help of good management on public perception which is created by the thoughts above important steps may occur on closure of energy deficit, economic benefits and more liveable nature. It should be expanded with the perception that can be handled as a rational way of eliminating the energy problem with a nuclear energy policy fulfilled by taking into people's political will. A national nuclear energy policy should be determined in the perspective of national and international developments as healthy and correctly and adopted a multi-dimensional nature and the technical management of the use of nuclear energy in accordance with the requirements. In addition to this, sample strategies must be arranged to make this policy happen. As a result, for the success of a strong nuclear energy program with the creation of a political and scientific infrastructure there should be a conscious and sincere commitment.

References

1. TAEA, Turkish Atomic Energy Authority Energy Report, (2010).
2. TUYAP, The Energy Sector in Turkey, Evgin Stock Shares Research Press, 1, Istanbul, February, (1998).
3. U. Gursoy, The Social Costs and Clean and Renewable Sources of Energy, Turkish Medical Association Publications, (2004).
4. R. Karluk, *Structural Transformation of Turkish Economy from 1923*, Beta Press, (2007).
5. A. Satman, Symposium of Energy and Development in Turkey: Energy Sources in the World, Tasam Press, (2006).
6. TUBITAK, Vision 2023 The Project of Technological Foresight, Panel of Energy and Natural Sources, (2003).
7. M. Parfit, *Alternative Energy*, National Geographic, August 2005, pp. 76-106, (2005).
8. S. Kocak and H. Altun, *Our Energy Needs and Nuclear Energy*, Symposium of New and Renewable Energy Sources, TMMOB, 3-4 October 2003, pp. 397-400, Kayseri, (2003).
9. M. Schneider and A. Froggatt, *The World Nuclear Industry Status Report*, Brussels, (2004).
10. Y. Cengel, *The Use of Geothermal, Wind and Other Renewable Energies in the World and*

Turkey", Symposium of New and Renewable Energy Sources, TMMOB, 3-4 October 2003, Kayseri, pp. 1-14, (2003).

11. E. Torunoglu, *Is it Ecological Disaster Threshold: and the World is Polluted*, Oteki Press, pp. 19-34. (1997).
12. Republic of Turkey Ministry of Energy and Natural Resources, Nuclear Energy Project Implementation Department Report, (2013).
13. V.S Ediger, *Energy The New World Order*, Turkish Sciences Academy, Academic Forum, No: 67, (2011).
14. BP, *Statistical Review of World Energy*, British Petroleum (BP), London, UK, (2012).

EPJ Web of Conferences **128**, 05002 (2016) DOI: 10.1051/epjconf/201612805002
TESNAT 2016

The importance of the different kinds of energy sources for energy future of Turkey

Yusuf Alper Kaplan[a], *Canan* Aladağ

Osmaniye Korkut Ata University, Energy System Department, Turkey

Abstract Nowadays, the need of energy has been increasing day by day with the population growth and the advancements of technology. In this study, the current state of nuclear, wind and solar energy on the worldwide has been generally investigated. The general assessments have been made based on Turkey's energy potential and the evaluation situation of this potential. The current political structures of countries are generally assessed and under this policy, the last situation and the latest implemented innovations are given. Turkey's energy demand is constantly increasing and Turkey is a country that needs to energy imports. This is a need for new energy sources to meet the growing need for energy. Nuclear, wind and solar energy are the new sources of energy to the fore in our country recently. In this study is given general information on the usage of energy sources of making and some deficiencies were been emphasized by political considerations in this regard.

1 Introduction

Energy sources gain in importance at a fast pace from the aspect of both, the energy requirements of the country and more so, environment. This work that aims to give, the status of the energy potential in Turkey and the current energy structures of Turkey's policy. Turkey has high potential in terms of many energy sources. Fig. 1 shows that Turkey's installed capacity reached to 73,854.6 MW level at the end of February 2016.

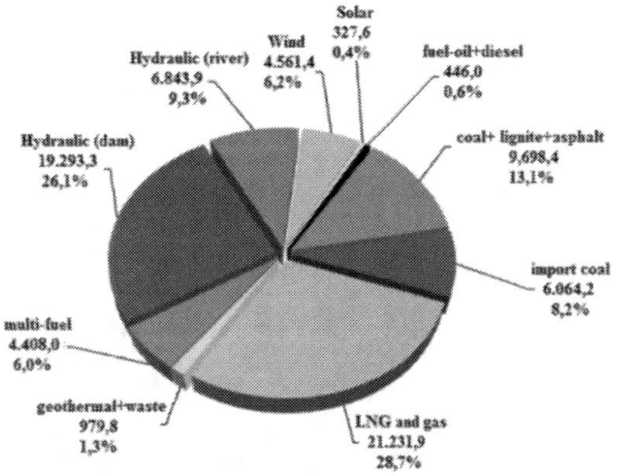

Fig. 1. Electric power installed capacity in Turkey (end of February 2016) [1]

The most important technique is to increase the diversity of energy sources to ensure energy security for the country. As it can be understood from the graph, use of nuclear energy does not have in Turkey. Significant investments are made in our country about nuclear, wind and solar energy sources and support policies to increase the use of them has been developed.

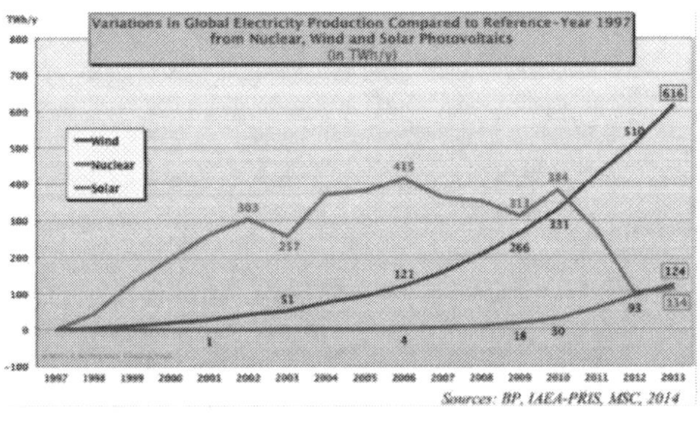

Fig. 2. Variation in Global Electricity Production from Nuclear, Wind and Solar [2]

Fig. 2 presents the actual electricity produced by solar photovoltaics, wind and nuclear power and highlights the changing levels of production since 1997[3].

2 Wind Energy

The importance of the using of renewable energy resources are realized while considering the limited duration of the use of fossil fuel resources. Wind energy sector has shown a rapid development all over the world in recent years. As a result of the last technological improvements, popular support mechanism and new incentives, the share in wind energy of power production has continuously increased.

[a] Corresponding author: yakaplan@osmaniye.edu.tr

DOI: 10.1051/epjconf/201612805002

2.1 The General State of Wind Energy in the World

The using of the wind energy varies temporally and geographically. The installed wind power capacity which is from 2004 to 2014 in the world is shown in Fig. 1. The total installed wind power capacity of the world was 238,050 GW with more than 20% growth rate in the annual market at the end of 2012. The wind power capacity in 2013 reached to 318,137 GW level with an increase of about 80 GW. At the end of 2014, the installed total wind power capacity reached 369,553 GW in the world [4].

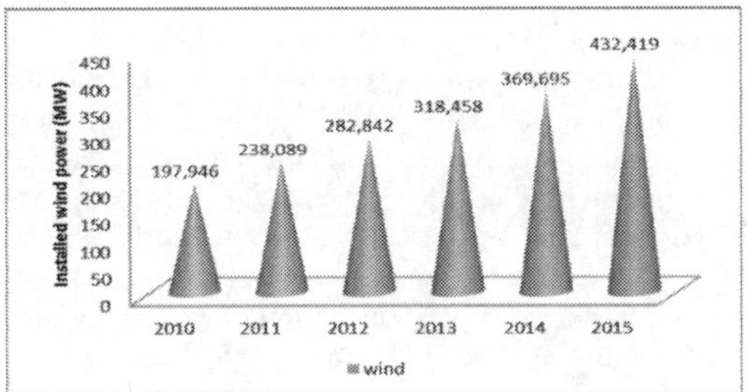

Fig. 3. Installed wind power capacity in the world [5]

2.2 The General State of Wind Energy in Turkey

Turkey is one of the richest countries in terms of wind power potential depending on its location. It is clearly seen in Fig. 4 that the installed wind energy capacity of Turkey seems to be very noticeable. It can be said that the total installed wind power capacity began to increase more rapidly last five years. The installed wind power capacity of Turkey is shown in Fig. 4 year by year. In 2012, wind power capacity is increased with 531 MW and total installed wind power capacity reached to 2260.5 MW level. In 2013, 499.1 MW of new wind power installed and total installed wind power reached 2759.6 MW level. Turkey added 58.2 MW of new wind power in 2015 for a total capacity of 4561.4 MW. It can be said that based on the Fig. 4 Turkey's total installed capacity has nearly %30 growth rate per year since 2011 and total installed wind power of Turkey 4561.4 MW in 2016.

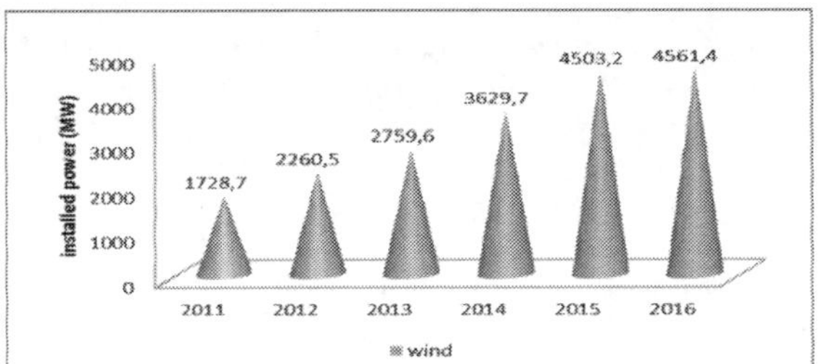

Fig. 4. Installed wind energy capacity in the Turkey

3 Solar Energy

Recently, the need of the using of energy has been increasing continuously with the improvements of technology and growth of the population for all countries. The prices of energy consumption in a country greatly affect its economy. In this case, the using of the renewable energy sources is a substantial research topic in the worldwide and its technology and investments improve progressively. Especially, EU countries are the pioneer of using renewable energy source instead of using fossil fuels [7]. Solar energy contributes diversification to using of the energy sources and this ensures the security of energy supplies. It also allows reclamation of degraded land and electrification of rural and remote communities [8].

3.1 Solar Energy in the World

The total installed solar energy capacity of the world from 2010 to 2014 is shown in Fig. 5. It is seen in the Fig. 5 that the world's total installed solar power capacity was 100 GW with more than 40% growth rate in the annual market at the end of 2012. The world's total installed solar energy capacity reached to approximately 138 GW at the end of 2013. This capacity is capable of generating as much annual electrical energy as 16 coal power plants. These PV installations save more than 53 million tons of CO_2 each year [9,10]. Nearly 39 GW of new solar power capacity was installed around the world in 2014, so the total installed capacity reach to 177 GW [9].

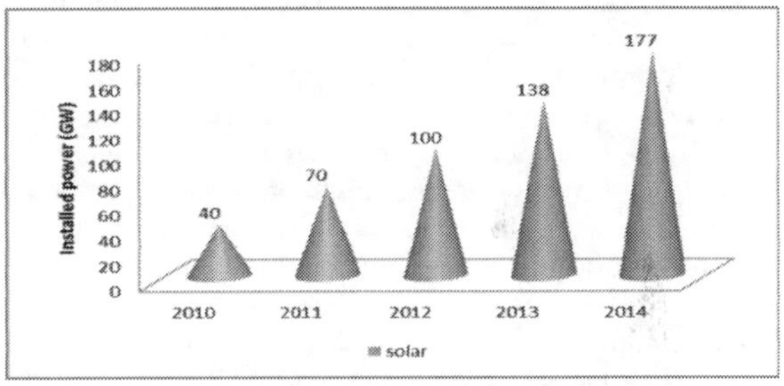

Fig. 5. Installed Solar Power Capacity in the World [11]

3.2 Solar Energy in Turkey

Turkey situated latitudes 36° and 42° north and due to its geographic position it is a fortunate country from the aspect of possessing a vast potential in solar energy, compared to quite a number of countries.

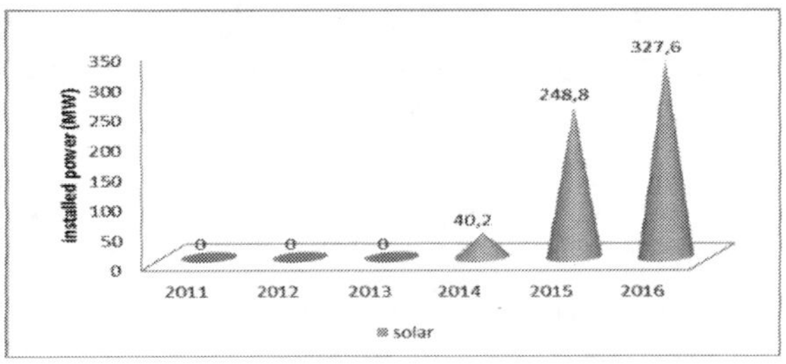

Fig. 6. Installed solar energy capacity in the Turkey [12]

The installed solar power in Turkey is shown in Fig. 6 year by year from 2011 until 2016. When you consider the potential of solar energy in Turkey, the result seems to be very dramatic. Turkey's solar energy potential have been identified that annual sunshine duration of Turkey is very remarkable but, it is clearly seen that the installed solar energy of Turkey begins from 2014. Installed solar power showed a significant increase from 2014 to today with 287.4 MW. Complicated and slow administrative procedures are the greatest obstacle for solar energy development in previous period in Turkey. The Turkish Solar Energy capacity has set a target of reaching 5 GW in 2023 [4]. There are some solar energy projects in Turkey such as "3000 MW Solar Farm Project" which will be constructed in Konya.

4 Nuclear energy

Nuclear is obtained as a result of reactions occurring in the core of the atom. With the implementation of an alternative energy development strategy, urgent strides are necessary to find sources of alternative energy. The generation of electricity from nuclear power plants with their lower-cost, greenhouse-gas-free clean energy and solid security should be an alternative in the world [13].

4.1 Nuclear Energy in the World

Nuclear energy is a mature low greenhouse gas emission source of base load power, but its share of global electricity generation has been declining for the past twenty years. While nuclear power could make an increasing contribution to low carbon energy supply, a variety of barriers and risks exist including operational risks, and associated concerns like the environmental impact of uranium mining, financial and regulatory risks, unresolved waste management issues, nuclear weapons proliferation concerns, and adverse public opinion [3].

But, nuclear energy have a big capacity in the energy of the world .the nuclear energy reactor and power capacity of some countries are given Fig. 7.

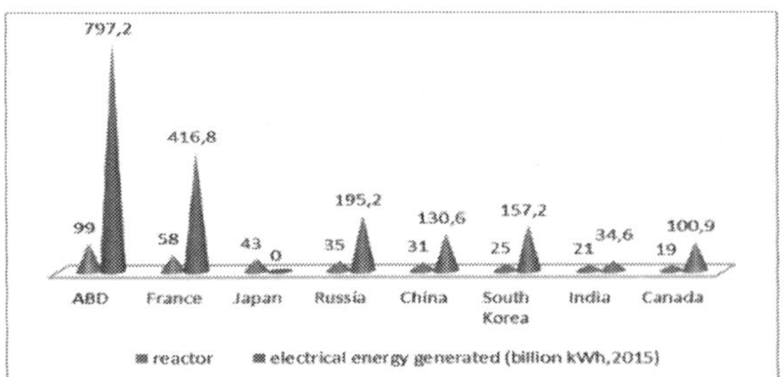

Fig. 7. Nuclear energy power and reactors for some countries [2]

4.2 Nuclear Energy in Turkey

Turkey, geographically deprived of natural resources, generate almost all of its electricity through imported energy sources such as natural gas and oil. To date, most of the electricity is generated through natural gas while the remaining is almost fully generated through oil. However, with the increasing concern about price volatility, rising electricity demand and energy security, Turkey is considering tapping into the field of nuclear energy to generate electricity. Nuclear energy, being a clean energy, also contributes toward environment sustainability by reducing the amount of carbon dioxide emitted into the atmosphere [14]. Nuclear power is often hailed as a technology which is important for mitigating climate change while providing energy necessary for economic development. Turkey is one of ten nuclear Newcomers which has most of the basic capacities and motivations historically observed in countries introducing nuclear energy: a large economy, an adequate electricity grid, and rapidly-growing electricity demand combined with energy dependence concerns. The Turkish state has always had a strong motivation for nuclear power. Since 1960 electricity consumption in Turkey grew on average 9% per year as compared to 7% for the world as a whole [15]. As a country poor in fossil resources, Turkey has been increasingly dependent on imported energy to meet its electricity needs and in recent decade energy security was cited as one of the main reasons for pursuing nuclear power. The establishment of planned nuclear power plants in Turkey are given in Fig. 8.

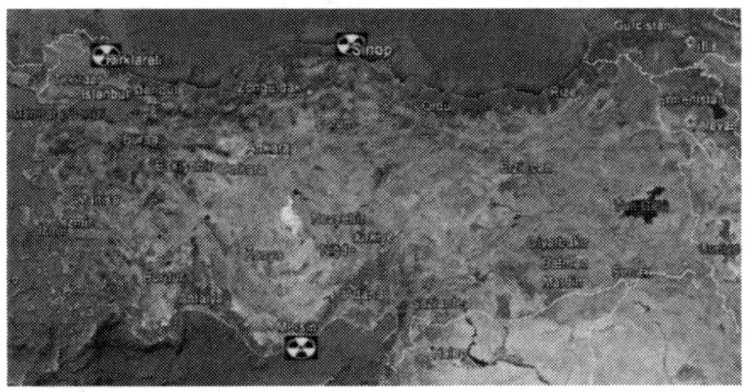

Fig. 8. The nuclear reactors are planned to be established in Turkey

In order to meet the fast rising demand for electricity, and reduce the risks associated with being dependent on imports, it is planned to take 2 nuclear power plants into operation by 2023, and to begin the construction of a further 3 during the same period. With this purpose in mind, the company, Akkuyu Nuclear Power Plant Electricity Generation, which was established within the framework of the agreement between the Republic of Turkey and the Russian Federation, which took effect on 27 December 2010, for the establishment of a nuclear power plant in the Mersin. Together with this agreement, it is envisaged that a nuclear plant comprising four VVER-1200 type units, and with a total installed capacity of 4.800 MW. On the other hand, an agreement was also signed in 2013, for the establishment of a nuclear plant in Sinop, and the work related to this is currently ongoing. That the power plant is expected to be 4480 MW [16].

Turkey Atomic Energy Agency (TAEK) according to the information received from the central resource for third-İğneada Kırklareli region forward as an alternative to leaving the focus on the Düzce-Akcakoca is stated.

Electricity transmission, distribution, due to advantages such as transportation voiced İğneada will be the first choice.

5 Expectations of Energy Future of Turkey

With a rapidly growing economy, Turkey has become one of the fastest growing energy markets in the world. Over the last decade, Turkey has been the second country, after China, in terms of natural gas and electricity demand increase. The limits of Turkey's domestic energy sources in light of its growing energy demand have resulted in dependency on energy imports, primarily of oil and gas. The primary aim of Turkey is to realize its own energy security. The fundamental objectives of energy policies in Turkey as followings:

- diversify its energy supply routes and source countries,
- increase the share of renewables and include the nuclear in its energy mix,
- take significant steps to increase energy efficiency,
- contribute to Europe's energy security.

In our country, to meet the increasing energy demand, to reduce dependence on foreign energy and environmental, energy production due to the positive impact of the decision was made to benefit from nuclear energy [17].

6 Conclusion

The current situation of nuclear, wind and solar energy in Turkey and the worldwide are generally evaluated by using recent data. This article also assessed the political structure of Turkey by looking at the installed wind and solar energy power. Turkey has rich wind and solar energy potential, which may be used for electricity power generation provided the necessary political and technological steps are taken. The rapid growth in renewable energy system installation due to the increase in energy demand had led to a growing demand for better politic structure as well as modelling. The most important deficiencies are determined and some proposals are presented in order to increase the using of renewable energy sources. Based on the supply and demand projections for electricity energy in our country, it is aimed that the share of the electricity generated by nuclear energy plants within the total, reach 5% by 2025.

The nuclear energy will be one of the most important energy source in the future for Turkey. Consequently, it is necessary to seriously studies on nuclear energy subject. Furthermore, universities should put more emphasis on high level education and training in the area of renewable energy.

References

1. www.**teias**.gov.tr
2. www.iaea.org Power Reactor Information System, (2016)
3. *Nuclear Power Versus Renewable Energy-Trend Analysis*, Vol. 103, No. 4, Proceedings of the IEEE, April (2015)
4. GWEC (Global Wind Energy Council), *Global wind 2014 report*, March, (2015)
5. www.ewea.org
6. www.tureb.com.tr
7. Y.A. Kaplan and I. San, Green Energy Conference-VI (IGEC-VI), Eskisehir, Turkey, (2011)
8. K.H. Solangi, et al., Renew Sustain Energy Rev, **15**: 2149–63, (2011)
9. *Renewable 2014 Global Status Report*, REN21, (2013)
10. European Photovoltaic Industry Association (EPIA), Global Market Outlook For Photovoltaics 2013-2017, (2013)
11. D. Bahrehmand, M. Ameri, Renewable Energy **74**, 357-368, 2015
12. http://dektmk.org.tr
13. A review on nuclear power plant scenario in Thailand, Parinya Pongsoi a, Somchai Wongwises, Renewable and Sustainable Energy, 586–592 (2013)
14. C.E. Seng, L.C. Kiat, *The System Dynamics of Nuclear Energy for Singapore*, IGEC-6, 5-9 June, (2011)
15. H.A. Simsek, N. Simsek, Recent incentives for renewable energy in Turkey, Energy Policy, **63**, 521–530, (2013)
16. http://enerji.gov.tr
17. www.mfa.gov.tr

 DOI: 10.1051/epjconf/201612806001

Estimation of monthly global solar radiation in the eastern Mediterranean region in Turkey by using artificial neural networks

Muhittin Sahan[1] and *Emre* Yakut[2]

[1]Osmaniye Korkut Ata University, Faculty of Arts and Science, Department of Physics, Osmaniye, Turkey
[2]The Faculty of Economics and Business Administration, Management Information Systems, Osmaniye, Turkey

Abstract. In this study, an artificial neural network (ANN) model was used to estimate monthly average global solar radiation on a horizontal surface for selected 5 locations in Mediterranean region for period of 18 years (1993-2010). Meteorological and geographical data were taken from Turkish State Meteorological Service. The ANN architecture designed is a feed-forward back-propagation model with one-hidden layer containing 21 neurons with hyperbolic tangent sigmoid as the transfer function and one output layer utilized a linear transfer function (purelin). The training algorithm used in ANN model was the Levenberg Marquand back propagation algorith (trainlm). Results obtained from ANN model were compared with measured meteorological values by using statistical methods. A correlation coefficient of 97.97 (~98%) was obtained with root mean square error (RMSE) of 0.852 MJ/m^2, mean square error (MSE) of 0.725 MJ/m^2, mean absolute bias error (MABE) 10.659MJ/m^2, and mean absolute percentage error (MAPE) of 4.8%. Results show good agreement between the estimated and measured values of global solar radiation. We suggest that the developed ANN model can be used to predict solar radiation another location and conditions.

1 Introduction

The Sun, which is a source of heat and light for all living things on Earth, is an average star with a diameter of $1.39x10^6$ km, with a distance of the earth of approximately $1.49x10^8$ km, and with an effective surface temperature of around 5800K [1]. It is known a source of the most important fusion process in nature. The sun is a natural nuclear fusion reactor in which hydrogen atoms combine to form helium and in the process release a tremendous amount of energy. Solar radiation on the Earth's surface varies from the solar radiation incident on the Earth's atmosphere. Latitude of a location and the time of the year cause variations in solar radiance at the Earth's surface. The spectrum of the solar radiation at the earth's surface has both direct radiation coming straight from the sun and diffuse radiation scattered from the sky and from the surroundings. The total ground radiation combining direct and diffuse radiation is called the solar global radiation.

Global solar radiation data are a fundamental input parameter for solar energy applications. Since solar radiation measurements are not easily available for many region, it is necessary to develop some methods to predict the solar radiation using available meteorological data. Many models such as empirical regression models or artificial neural network (ANN) models have been developed to predict the amount of global solar radiation on horizontal surfaces using various climatic parameters, such as relative humidity, sunshine duration etc. It is known that artificial neural networks (ANNs) models discussed in the literature are shown to be more suitable to predict solar radiation than other empirical regression models [2,3]. The ANN models are generally presented as systems of interconnected neurons and are widely used as an alternative way to solve many complex problems. The ANN models provide a computationally efficient way of determining an empirical, possibly nonlinear relationship between a number of inputs (such as latitude, longitude, humidity, sunshine duration) and one or more outputs (such as solar energy). ANN models are type of solar prediction models and there have been several articles that have used artificial neural networks for predicting solar radiation [4-7].

2 Data Description and Methodology

A total five metrological stations with long-term available data of global solar radiation were used in this study. The geographical information (longitude, latitude, and altitude) for these stations are listed in Table 1. The stations roughly range from 36°.15 to 37°.26 latitude North and from 35°.19 to 36°.10 longitude East and from 2 to 1344m altitude (From mean sea Level).

Table 1. Geographical features of five stations from Mediterranean region (Turkey).

Location	Latitude (°N)	Longitude (°E)	Altitude (m)
Adana	37.00	35.19	27
Kozan	37.26	35.49	115
Karata	36.34	35.23	22
Antakya	36.15	36.08	100
Iskenderun	36.35	36.10	2

In the present study, we developed a computer program running under MATLAB (R2015a) using Neural Network Fitting Tool (nftool). Artificial Neural Network (ANN) was used to analyses the monthly global solar radiation data belonging to 5 centers of localization (Table 1) from east Mediterranean region in Turkey. Data from 5 stations averaged over eighteen years (1993-2010) were used to train and test the networks. The data sets of 80% were used for ANN training and the remaining 20% data sets were used for ANN testing. Once the data are trained, they can perform predictions at very high speed [7].

A typical ANN consists of an input layer, weights, a hidden layer, a transfer function, and output layer as shown in Fig. 1. Fig. 1 shows a basic diagram of an artificial neuron in a feed forward network in ANN. The input layer and the weight at the inputs consist of n elements (xn,wn). The hidden layers, considered as the second layer, contain a nonlinear transfer function such as tansig or logsig. The output layer has a linear transfer function such as purelin. Each input element is multiplied by the relevant weight and then these products and bias are collected with a summing function. The result of summation is sent to the activation function to obtain output data [8].

In order to estimate the global solar radiation, we used well known Multi-Layer Perceptron (MLP) Network and we selected a multilayer feed-forward back-propagation artificial neuron network (FFBPANN) model for ANN structure. This FFBPANN means that the artificial neurons are organized in layers, and send their signals forward, and then the errors are propagated backwards [7].

In order to determine the ANN network model, various models with different training algorithms and architectures (approximately 100 models) have been employed and tested during the training of our network. Diagnostic linear regression statistic, including R^2, RMSE, MSE, MBE, MABE, and MAPE (see following paragraph) have been used in order to determine the best ANN model. For this reason, we used the highest values of R^2 and the lowest values of RMSE, MSE, MBE, MABE, and MAPE. Some of these employed and tested training algorithm models are given in Table 2.

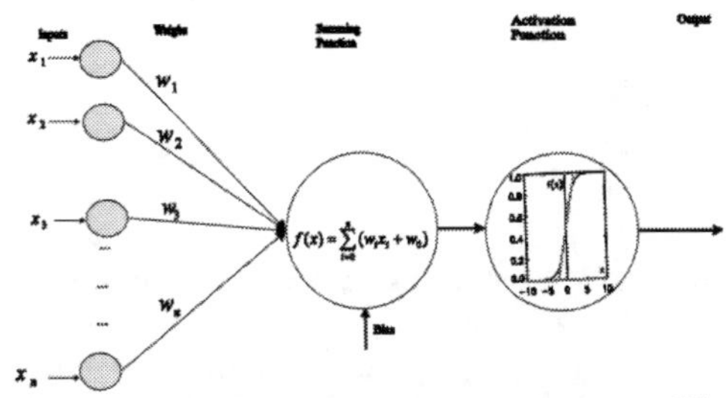

Fig. 1. Caption of the Figure 1. Below the figure. The basic diagram of a neuron in a feed forward network in ANN [8].

Table 2. Various ANN models developed with different training algorithms and architectures. (Training Algorithm: Tr.Alg., Hidden transfer Function: H.Tr.F, Output transfer Function: O.Tr.F, Number of neurons in hidden layer: Num.N.H.L)

Tr.Alg.	H.Tr.F	O.Tr.F	Num.N.H.L	R^2	RMSE	MSE	MABE	MAPE
trainlm	tansi	purelin	21	0.98	0.85	0.72	0.65	0.04
trainlm	tansi	purelin	5	0.97	0.85	0.72	0.62	0.04
trainlm	tansi	tansig	5	0.97	0.89	0.79	0.69	0.05
trainlm	logsi	tansig	33	0.97	0.85	0.72	0.68	0.05
trainlm	logsi	tansig	30	0.97	0.92	0.84	0.71	0.05
trainlm	tansi	purelin	12	0.97	0.93	0.86	0.71	0.05
trainlm	tansi	purelin	18	0.97	0.94	0.90	0.74	0.05
trainbf	logsi	tansig	20	0.97	0.94	0.88	0.71	0.05
trainlm	logsi	tansig	24	0.97	0.99	0.98	0.70	0.05
trainbf	logsi	tansig	10	0.97	1.01	1.02	0.76	0.05
trainbf	logsi	tansig	35	0.96	1.01	1.03	0.70	0.05
trainsc	logsi	purelin	8	0.95	1.24	1.55	0.95	0.08
trainlm	tansi	purelin	33	0.97	0.90	0.81	0.69	0.05
trainlm	tansi	purelin	39	0.97	0.92	0.85	0.70	0.05
trainlm	tansi	purelin	36	0.97	0.92	0.85	0.70	0.05
trainlm	tansi	tansig	30	0.97	0.94	0.90	0.73	0.05
trainlm	logsi	purelin	12	0.97	0.90	0.81	0.70	0.05
trainbf	logsi	tansig	15	0.97	0.90	0.81	0.70	0.05
trainlm	tansi	tansig	27	0.97	0.93	0.88	0.73	0.05
trainlm	logsi	tansig	18	0.97	0.94	0.88	0.73	0.05
trainlm	tansi	tansig	33	0.97	0.95	0.90	0.73	0.05
trainlm	tansi	purelin	15	0.97	0.94	0.88	0.73	0.05
trainlm	logsi	purelin	10	0.97	0.93	0.87	0.73	0.05
trainlm	logsi	purelin	21	0.97	1.00	1.00	0.75	0.05
trainlm	tansi	tansig	36	0.97	0.92	0.86	0.68	0.04
trainlm	logsi	purelin	39	0.97	0.95	0.91	0.73	0.05

The best models developed according to training algorithms and transfer functions, and number of the neurons in the hidden layer have been selected. The "trainlm" algorithm was used in order to train the ANN network and "tansig", and "purelin" activation functions (see Fig. 2) were used in the hidden layer and in the output layer, respectively. Our ANN architecture consists of three layers called as input layer, hidden layer and output layer. In these layers, we used 7 input neurons, 21 hidden neurons which are the collection of neurons, and a single output neuron (namely, 7-21-1) to be able to generate a modest instance for training and then testing our data. Figure 2depicts the activation function used in the known Multi-Layer Perceptron forecaster network.

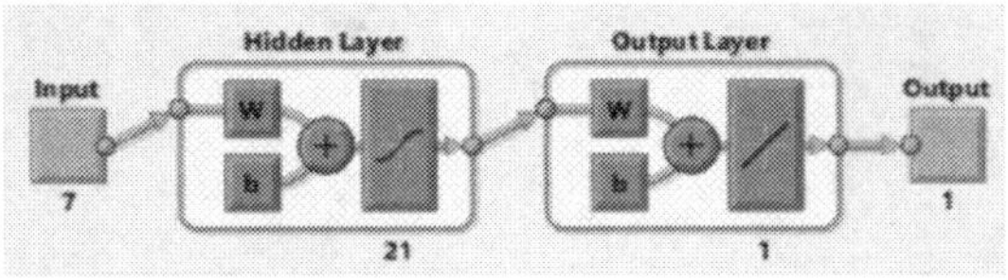

Fig. 2. Activation function used in Multi-Layer Perceptron network.

There are several statistical measures available to evaluate the association between estimated and measured values [9]. The famous of them can be given as the correlation coefficient of determination (R^2), the RMSE (Root Mean Square Error), the Mean Absolute Percentage Error (MAPE), Mean Square Error (MSE), and MABE (Mean Absolute Bias Error). These statistical indicators are defined by the following mathematical equations:

$$R=\frac{\sum_{i=1}^{n}(x_i-\bar{x}_i)(y_i-\bar{y}_i)}{\sqrt{\sum_{i=1}^{n}(x_i-\bar{x}_i)}\sqrt{\sum_{i=1}^{n}(_i-\bar{y}_i)}}, \qquad MAPE=\frac{1}{n}\sum_{i=1}^{n}\left|\frac{(x_i-y_i)}{x_i}\right|\times 100$$

$$RMSE=\sqrt{\frac{1}{n}\sum_{i=1}^{n}(x_i-y_i)^2}, \qquad MSE=\frac{1}{n}\sum_{i=1}^{n}(x_i-y_i)^2 \qquad (1)$$

$$MABE=\frac{1}{n}\sum_{i=1}^{n}|x_i-y_i|$$

where y_i and x_i parameters represent the estimated and the measured (actual) values, respectively and n is the number of data points. The correlation coefficient of determination (R^2) is used to determine the linear relationship between the estimated and measured values. The R^2 varying from 0 to 1 is adopted to measure the fit of the model on estimated data. A model is more efficient when R^2 is closer to 1 [10]. The value of RMSE provides information on the short term performance of the correlations and is always positive and zero in the ideal case.

3 Results and Discussions

In this study, we applied ANN techniques to predict monthly global solar radiation using meteorological and geographical data from five stations from eastern Mediterranean region (Turkey) as Adana, Kozan, Karatas, Antakya, and Iskenderun (see Table 1). Meteorological and geographical data include the monthly mean values of the global solar radiation on horizontal surface, air temperature, relative humidity, sunshine duration, latitude, longitude, month, and altitude. All meteorological and geographical data between 1993-2010 (18 years) were obtained from Turkish State Meteorological Service (TSMS).

The actual performance of the ANN models during training and testing procedure are obtained using the basis of statistical error tests given at Eq. 1. It is found that the most successful network was single layer network with 21 neurons in hidden layer as mentioned in pervious section. Fig. 3 shows typical ANN used for estimating of solar radiation. As seen from Fig. 3, while four geographical parameters (month, latitude, longitude, and altitude) and three meteorological parameters (monthly average air temperature, relative humidity, and sunshine duration) are used as input layer, solar radiation values are obtained from the output layer (see Fig. 3).

Comparison between measured and estimated monthly global solar radiation values for all selected regions from Mediterranean region in Turkey were given in Fig. 4. Data points (with symbol filled circle) given in Fig. 4 represent measured and estimated monthly global solar radiation data. A regression line was fitted to monthly estimated and measured global solar energy values in order to compare with each other as seen in Fig. 4. Regression equation and R^2 values are labeled in the upper left corner of Fig. 4. y and x parameters in regression equation represent measured and estimated solar energy, respectively. As seen from Fig. 4, the estimated solar radiation from the ANN model were correlated with the measured values, giving coefficient of determination (R^2) of 98%. This means that there is a good fitting between the estimated using ANN model and measured monthly mean global solar irradiation.

Monthly global solar radiation estimations from the ANN model were compared with the actual data by using simple error analysis; R^2, RMSE, MSE, MABE, and MAPE to the monthly mean value for the five variables under study. Results for each location were given in Table 3. Table 3 contains results of R^2, RMSE, MAPE MSE, and MABE values for selected five locations. As seen from Table 3, the linear correlation coefficients (R^2) are high for all the variables changing between a minimum of 97.1% at Adana and a maximum of 99.5% at Karatas. This implies that there are statistically significant relationships between estimated actual measured global solar radiation. The lowest RMSE value is found for location of Kozan (0.506 MJ/m^2) and the maximum for Karatas (1.110 MJ/m^2). The MAPE estimate for the five locations ranges between 2.8% for location of Kozan and 6.4% for location of Karatas with an overall average of 0.048%. The highest and lowest MSE values were calculated as 1.231MJ m^{-2} for Karatas and 0.256MJ m^{-2} for Kozan. The highest and lowest MABE values were calculated as 0.941MJ m^{-2} for Karatas and 0.369MJ m^{-2} for Kozan.

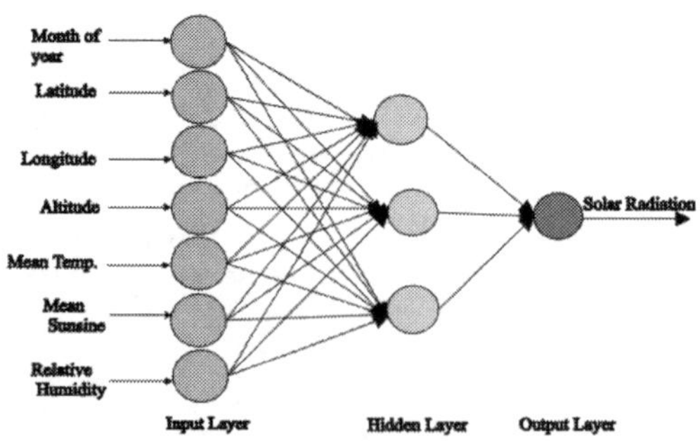

Fig. 3. Activation function used in Multi-Layer Perceptron network.

Results obtained from ANN model were compared with measured meteorological values by using statistical

DOI: 10.1051/epjconf/201612806001

methods. Fig. 5 presents monthly comparison between estimated from ANN and measured values of solar radiation for five locations (Adana, Kozan, Karataş, Antakya, and Iskenderun) given in Table 3. Location names are labeled in the upper left corner of each plot. Measured and estimated values are represented with open and filled circles and are labeled in the upper left corner of each plot. As seen from Fig. 5 and Table 3, the best R^2 were formed for locations of Karatas (0.995%) and Kozan (0.994%).

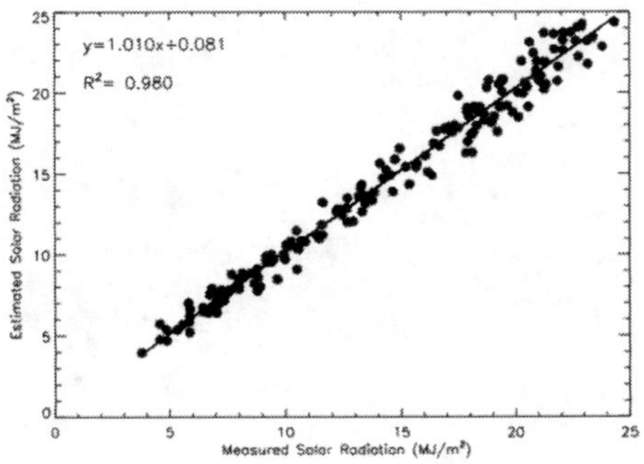

Fig. 4. Comparison between measured and estimated monthly global solar radiation values of the correlation equation for all regions from Mediterranean region in Turkey.

Table 3. Comparison of the statistical errors between measured and estimated monthly global solar radiation for five stations.

Location	R^2 (%)	RMSE (MJ/m^2)	MSE (MJ/m^2)	MABE (MJ/m^2)	MAPE (%)
Adana	0.971	1.007	1.013	0.814	0.056
Kozan	0.994	0.506	0.256	0.369	0.028
Karatas	0.995	1.110	1.231	0.941	0.064
Antakya	0.991	0.601	0.361	0.472	0.045
Iskenderun	0.992	0.874	0.764	0.696	0.047

4 Conclusion

Meteorological Service (TSMS). Results obtained from ANN model were compared with measured meteorological values by using statistical methods. These results show a good agreement between the measured by ANN model and estimated monthly mean global solar radiation values. We propose that the ANN model can be applied to estimate global solar radiation for any region having meteorological and geographical parameters.

In this study, an artificial neural network based model for prediction of solar energy has been developed. Meteorological data of five meteorological stations in Mediterranean region (Turkey) for period of 18 years (1993–2010) were taken from Turkish State Meteorological Service (TSMS).

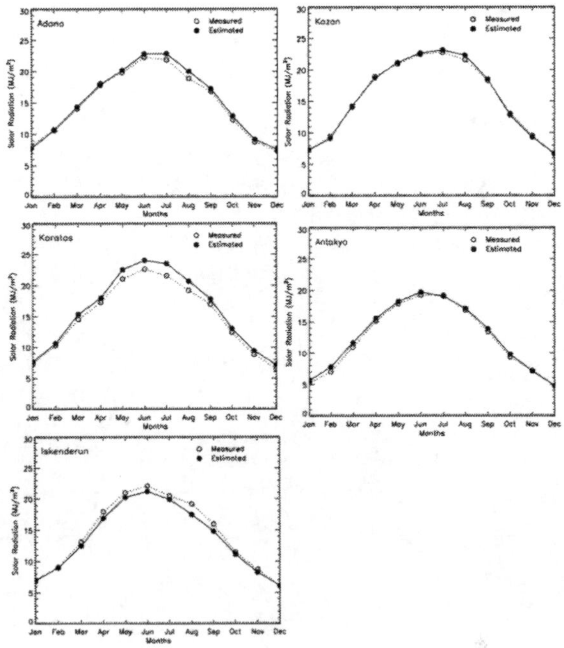

Fig. 5. Monthly mean variation of estimated (by ANN model) and measured values of global solar radiation for five different stations selected from Mediterranean region in Turkey.

References

1. J.A. Duffie and W.A. Beckman, "Solar Engineering of Thermal Processes, 2th Edition", John Wiley and Sons Inc., New York, (1991)
2. D.A. Fadare, Applied Energy, **86** (2009)
3. Y. Jiang, Energy Policy, **36** (2008)
4. M.A. Abdul Azeez, Scholars Research Library, Archives of Applied Science Research, 3, 2 (2011)
5. A. Moreno, M.A. Gilabert, B. Martinez, Solar Energy, **85** (2011)
6. E. A Ahmed, M. E. Adam, Journal of Clean Energy Technologies, 1, **2** (2013)
7. S. Edalati, M. Ameri, M. Iranmanesh, Journal of Renewable Energy and Environment, 2, **1** (2015)
8. D. Graupe, Advanced Series in Circuits and Systems, **6** (2007)
9. E. Yakut, B. Elmas and S. Yavuz, Suleyman Demirel University The Journal of Faculty of Economics and Administrative Sciences, 19, **1** (2014)
10. S.A. Kalogriou, Appl Energy **67** (2000)

AUTHOR INDEX

EDP Sciences – Web of Conferences
17, Avenue du Hoggar
Parc d'Activité de Courtabouf
BP 112
F-91944 Les Ulis Cedex A
France

ISBN 978-1-5108-3283-1

9 781510 832831

CURRAN ASSOCIATES INC.
proceedings
.com